Mechanical Reliability

Oleksandr Grynchenko • Oleksiy Alfyorov

Mechanical Reliability

Prediction and Management Under Extreme Load Conditions

Oleksandr Grynchenko
Vasylenko National Technical University
of Agriculture
Kharkov, Ukraine

Oleksiy Alfyorov
Vasylenko National Technical University
of Agriculture
Kharkov, Ukraine

ISBN 978-3-030-41566-2 ISBN 978-3-030-41564-8 (eBook)
https://doi.org/10.1007/978-3-030-41564-8

This Springer imprint is published by the registered company Springer Nature Switzerland AG
The registered company address is: Gewerbestrasse 11, 6330 Cham, Switzerland

Preface

Recommended for publication by the Academic Council of Kharkiv Petro Vasylenko National Technical University of Agriculture, Protocol No. 10th April 27, 2017.

Reviewers: Doctor of Technical Sciences, V.V. Aulin (Central Ukrainian National Technical University, Kropivnitsky, Ukraine); Doctor of Technical Sciences, O.O. Larin (National Technical University "Kharkiv Polytechnic Institute")

Grynchenko O.S. Alfyorov O.I. *Fundamentals of Prediction and Management of Reliability Under Extreme Loads Conditions*—Kharkiv, 2019

ISBN UDC 62-192(075)

The monograph describes the main theoretical propositions of the methodology to predict mechanical reliability under conditions of repeated exposure to random extreme loads. The mechanical load process is considered to be a form of a discrete sequence of loads occurring at times that form a random flow. It is solved some problems of reliability prediction of elements having deterministic or random limit load. A method for the probabilistic justification of safety factors has been developed, providing a predetermined level of reliability of elements and systems for sudden failures when designing. It is considered the methods of prediction and managing reliability under conditions of using safety devices. The main theoretical results are presented in a form available for practical engineering applications. The book can be used as a manual by teachers and graduate students of higher technical educational institutions.

Гринченко О.С. Алфьоров О.І. Основи прогнозування та керування надійністю
в умовах екстремальних навантажень - Х.: ТОВ "Планета-Принт", 2017. - 136с.

ISBN 978-617-7587-22-3
УДК 62-192(075)

В монографії викладені основні теоретичні положення методології прогнозування механічної надійності в умовах багаторазової дії випадкових екстремальних навантажень. Процес механічного навантаження розглядається, як дискретна випадкова послідовність навантажень, виникаючих у моменти часу, що утворюють випадковий потік. Розв'язуються задачі прогнозування імовірності безвідмовної роботи елементів, які мають детерміновану або випадкову несівну здатність. Розвинуто метод імовірнісного обґрунтовування коефіцієнтів запасу, які забезпечують проектувальний рівень надійності елементів та систем відносно раптових відмов. Розглядаються способи прогнозування та керування надійністю в умовах використання запобіжних пристроїв. Основні теоретичні результати викладено у формі, яка є доступною для практичного інженерного застосування. Книгу можливо використовувати у якості посібника викладачами і аспірантами вищих технічних навчальних закладів.

ISBN 978-617-7587-22-3

Kharkiv, Ukraine

O. S. Grynchenko
O. I. Alfyorov

2019

Introduction

Reliability prediction of machinery parts being designed is primarily aimed at avoiding mechanical failures in service, caused by the influence of force, mechanical interaction of parts among themselves and with the environment. Special attention should be given to sudden mechanical failures due to the fact that their development, unlike gradual failures, as a rule, does not lend itself to individual diagnosis and prevention. In addition, as practice shows, sudden failures can often occur in the initial operation period of the equipment, which adversely affects its competitiveness when it is introduced to the market.

Most transport, agricultural, road-building, and other mobile machines have operating conditions, use modes, and, therefore, modes of mechanical loads of elements with a wide range of variation. Most of the time, a machine normally operates under normal (nominal) load conditions, whereby a long-term upstate and durability should be provided. However, in some relatively rare cases, some parts of machines experience short-term and repeated effects of extreme loads close to breaking or inelastic material deforming material details. The concept of "extremality," conditionally and in the following, actually means going beyond the normal and long-term stress mode. The degree of such an exit in practice should be further evaluated in each specific case. Elements subjected to the risk of sudden abandonment of repeated extreme overloads should be identified in the design of a special group in order to ensure their reliability.

The term "extreme loads" is used further in the sense that we consider the peaks of various mechanically loaded factors or combinations; they are potentially capable, even as a result of a single load that causes a sudden loss of the limit load by an element. Modern computer methods for analyzing the stress–strain state of machinery elements of different geometric shapes allow the engineer to estimate loads and stresses efficiently; they can lead to losses of the limit load in the element in accordance with the accepted design load scheme. However, this deterministic approach is insufficient to predict reliability. It is necessary to take into account that under the operation conditions the values of acting extreme loads have significant random dispersion, and the random variability of the mechanical characteristics of materials and technological modes of manufacturing elements together lead to the randomness of the value of the limit load.

The book is devoted to a brief review of the fundamentals of one of the many directions of science of reliability—predicting risks of sudden mechanical failures and associated reliability indicators. Theoretically, there are also promising ways to manage and ensure reliability under repeated extreme loads. They are based on stochastic models of emerging and preventing from sudden mechanical failures, typical for parts of mobile machines. It is possible to use the methods described when predicting reliability of building structures. The conceptual feature of setting tasks of mechanical reliability is to ensure the limit load of structures in time.

A probabilistic approach to reliability prediction is widely used, which corresponds to the stochastic nature of the influencing factors and causes of sudden mechanical failures. The monograph develops a method of probabilistic substantiation of the value of the safety factor under repeated loads, which refers to the ratio of the average values of the limit load of elements and extreme loads. This allows to carry out such a substantiation objectively on the basis of studying and analyzing test results and real statistical data on loads and on dissipating the limit load of structural elements. This way seems to be more progressive than the method of expert appraisal of safety factors that has been used up to now; which normative documents commonly used in designing in various sectors and departments are based on it.

The safety factor is a generalized parameter that in many respects determines the future material consumption and the cost price of the designed product. Therefore, dependencies linking the safety factors of the machine parts with the predicted reliability indexes allow, at the design stage, to choose a rational and economically feasible option to ensure a sufficient level of reliability. The same concerns justification for using safety devices that provide mechanical reliability. Therefore, the use of a set of methods for forecasting and managing reliability stated in the book should improve the efficiency of work related to the design of engineering products.

In addition to theoretical results and conclusions that are of general importance, the monograph also contains information that allows solving applied problems, carrying out practical calculations, and the engineering analysis of reliability of elements and systems. In particular, using the reference data in the tables of the appendixes to the main text, it is possible to predict an approximate value of the gamma-percentile operating time to a sudden mechanical failure that corresponds to a known value of safety factors and guarantees a given reliability function of the projected object. This takes into account the possible level of random dispersion of extreme loads and limit load of the elements.

The authors are grateful to the colleagues from the department of Reliability, Strength, and Technical Service of Machines named after V.Ya. Anilovich, who assisted in the preparation of the monograph. This is especially true to L.V. Blagutina, who professionally performed a computer suite of the repeatedly refined and formulas-rich text. Many computer calculations were performed by A.P. Yurieva. V.B. Savchenko helped a lot in finalizing and preparing the book for publication.

Contents

Chapter 1
Reliability of Elements with Deterministic Limit Load

1.1 Sudden Failures and Reliability Under Extreme Loads

Sudden mechanical failures of elements and systems in mobile machines are mainly due to repeatedly random extreme loads, which can jump at least once exceed the load capacity and lead to quasistatic destruction or occurrence of unacceptable residual deformations. The limit load of each element for this type of failure should also be considered as a random variable and therefore the patterns of sudden mechanical failures are stochastic. An essential feature of the model under consideration to predict a sudden mechanical failure is that its risk is not associated with accumulation of damage and does not depend on the history of loads; it does not affect the limit load.

Ensuring the reliability of machinery elements from sudden mechanical failures under the traditional deterministic approach to design calculations is reduced to the use of safety factors [1]. Their purpose is to take into account and compensate the influence of various random factors on mechanical characteristics of materials and the magnitude of operating loads. Thus, the influence of randomness when reliability is being provided is actually recognized, however, probabilistic models and methods are not used to account this influence. The safety factor is specified in the technological normative documents or standards used in various branches of engineering. Their value is usually established empirically on the basis of expert analysis and generalization of previous experience in the design and operation of products of a similar purpose.

As a rule, with this approach, it is impossible for an engineer to determine which failure risk of the projected object corresponds to the recommended standard value of the safety factor. He does not receive information about how the operation time of an object, its structure and the number of elements it consists of, affect failure risks. In connection with these drawbacks, the use of stochastic (probabilistic and statistical) concepts and corresponding models is more progressive when predicting

O. Grynchenko, O. Alfyorov, *Mechanical Reliability*,
https://doi.org/10.1007/978-3-030-41564-8_1

reliability. We should point out a number of advantages of the probabilistic approach to justification of the safety factors [2, 29, 30, 31, 32] over deterministic ones.

Firstly, using probabilistic models, it is possible to directly relate a value of the standardized safety factor to the predicted probability of non-destructive failure (reliability function) or the destruction risk expressed quantitatively and having a practical and practical meaning. The probability of no-failure operation is convenient from the point of rationing view, since it allows us to statistically evaluate possible material damage in case of a failure and justify the established standards economically.

Secondly, application of probabilistic models must be accompanied by additional information on characteristics of random scattering of the limit load of the elements and operating extreme loads, i.e. loads that are potentially capable of leading to sudden destruction or unacceptable deformations. Reasonable specification of these characteristics is possible on the basis of the statistical analysis of the results of mechanical tests of the materials used and field tests of machine elements. Real operational loads can also be objectively studied in a statistical way. This information as a whole contributes to a more rational and objective solution to the problems of ensuring the mechanical reliability of machines being designed.

The third feature to be noted is a possibility of taking into account the effect on the probability of no-failure operation of the number of repeated extreme loads. This allows us to relate a normative value of the safety factor to the expected number of random loads or the amount of operating time that must be ensured with a given probability of no-failure operation. An important advantage is a possibility of establishing on a probabilistic basis the relationship between the structure and the number of elements in the projected facility and requirements for the mechanical reliability of each element.

When predicting the mechanical reliability of machinery elements, two approaches to load estimation and reliability indicators are possible and used in practice. If static or dynamic calculations of the mode of deformation are carried out by analytical or numerical methods (finite element method or super element method), then at each point (finite element) of the calculated part under a given design scheme and the set of external influences, the tensor of the stress state components becomes known any of the strength hypotheses [3, 4] the only one parameter is evaluated - an equivalent voltage, the magnitude of which determines the hazard (risk) of destruction. Thus, for example, when using the Coulomb law, the equivalent stress is determined by the formula

$$\sigma_e = \sigma_1 - \eta \cdot \sigma_3$$

where σ_1 and σ_3—the greatest and smallest principal stresses; $\eta = {}^{\sigma_u}/_{\sigma_c}$—ratio of ultimate stresses for a material under uniaxial tension and compression.

When using the Mises hypothesis, the stress intensity is considered as the equivalent stress determined by the main stresses $\sigma_1 > \sigma_2 > \sigma_3$ according to the formula:

$$\sigma_e = \sigma_i = \frac{1}{\sqrt{2}}\sqrt{(\sigma_1 - \sigma_2)^2 + (\sigma_1 - \sigma_3)^2 + (\sigma_2 - \sigma_3)^2}$$

The hypotheses of Coulomb and Mises are more in line with the experimental data on the testing of ductile materials. A certain generalization of the experimental data allowing calculations of both ductile and brittle materials is given by the Pisarenko-Lebedev hypothesis, according to which the equivalent voltage is determined by the expression

$$\sigma_e = \eta \cdot \sigma_i + (1 - \eta) \cdot \sigma_1.$$

A detailed consideration of various strength hypotheses with the analysis of their applicability in engineering calculations is available in [4].

The reliability analysis is then carried out by probabilistic comparison of the possible values of equivalent stresses with the corresponding limiting characteristics of the material of the part, obtained usually under uniaxial tension and taking into account the features of the manufacturing technology.

The second approach finds application when the predictive estimation of mechanical reliability is carried out directly on the basis of results of full-scale tests of units and machinery and measurements of operational influences on them. Then, when predicting reliability, it is more convenient to compare the absolute extreme values of the main parameters of the external loads of the element (torque, dynamic concentrated and distributed loads, vibration acceleration, etc.) with the corresponding limiting (destroying) values of these same factors, determined experimentally or with the help of calculations.

In the following, the term "load" will be considered as one generalized parameter due to external loads; it completely determines the risk of sudden mechanical failures of an element or system. Limit load is a limiting value of the same parameter for an element. If during the object life cycle at any time of stresses, the "limit load" exceeds at least once, a sudden mechanical failure will occur at that moment, almost independently from the prehistory of loads.

Many elements of mobile machines operate under load conditions characterized by the presence of two components: main (normal) and additional extreme (overload). The normal variable component of the load is usually a random stationary process and is caused, as a rule, by steady dynamic processes of oscillatory nature.

However, in addition to the main stationary operation modes, the elements of mobile machines can additionally be subjected to short-term extreme actions that are of non-stationary nature. The time of their occurrence and magnitude are random and practically unrelated to the main stationary mode and its characteristics. Such areas are connected with transient dynamic processes in transmission, suspension, other elastic systems and their elements. They are non-stationary and superimposed on the stationary process. They can be caused by acceleration (starting from the place) or sudden braking, a sudden impact of the wheel on a separate unevenness or falling into a pothole; foreign bodies in the treated area, etc.

The range of extreme loads that arise in this case can be estimated by calculation methods [5, 6, 33, 34], and also on the basis of special field and bench tests [7, 8, 35]. The number of such repeated extreme loads for the entire life cycle of a particular element is random and can be quite significant, calculated in tens and hundreds. At the same time, recording and statistical analysis of the real values and frequencies of extreme loads of this type in the elements of mobile machines are usually difficult because they are rare and short-lived in operating conditions. In practice, sometimes the possibility of occurrence of such loads is revealed only by their consequences - to cases of sudden destruction or deformation of elements. In these conditions, the methods of objective inverse analysis of the operational data on the mechanical reliability of constructively similar analogues-predecessors acquire special significance in the design.

Practical methods of managing and ensuring the mechanical reliability at the design stage of machine elements under repeated exposure to accidental extreme loads are as follows:

- Designing with sufficient and reasonably probable safety factors, taking into account the random dispersion of the limit load and repeatedly acting extreme loads (materials, shape and size of parts, their rational arrangement in the nodes, etc.);
- Increasing the lower limit of dispersion of the limit load of the elements due to their preloading or the use of other methods of continuous monitoring of the lower resistance level to the sudden failure.
- Limiting the maximum loads on a part by applying safety clutches, strain limiters, safety valves and other devices that protect parts from overloads.

1.2 Stochastic Models of Reliability Under Multiple Random Loads

Schematization of extreme load and sudden failure. Let us consider applied stochastic models for predicting the probability of uptime, which allow us to perform engineering analysis and choose a rational alternative to ensure the required level of reliability.

Figure 1.1 shows the scheme of superposition of additional non-stationary extreme loads on the main stationary mode. Extreme load is a load potentially capable of leading an element to sudden destruction or unacceptable deformation. In practice, this means that the extreme load may exceed the lower limit of the random dispersion of the limit load. A value of the random extreme load $P_{\text{н}}$ is determined by the difference between the maximum of the extreme load and the lower limit of the distribution of the limit load, $\widetilde{P}_o$ which is a conditional non-damaging level. In practice, such a level can be taken, estimating it indirectly, as the upper limit of the range of possible values of the main constantly operating stationary loading process. For example, using the known "three sigma rule", we can

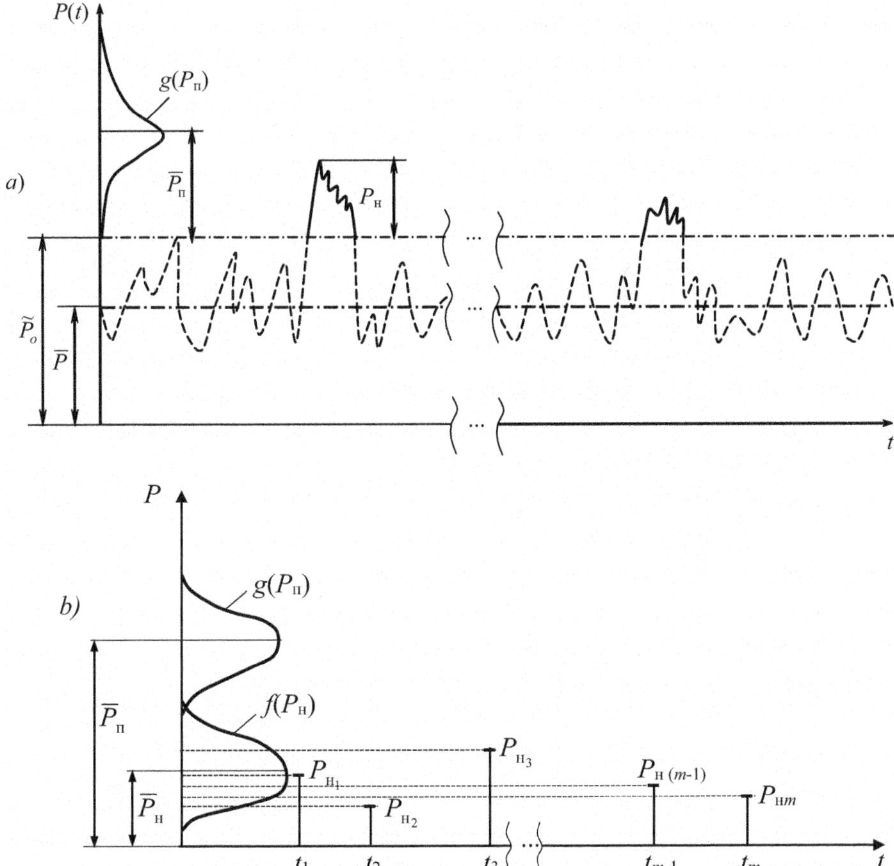

Fig. 1.1 Schematization of multiple extreme loads: (**a**) imposition of extreme loads on the stationary mode; (**b**) a scheme of a random stationary flow of independent discrete loads. Figure (**b**) shows: $g(P_{\Pi})$ distribution density of the random limit load; $f(P_{H})$ distribution density of random extreme loads

assume that $\widetilde{P}_o = \overline{P} + 3\sigma_\nu$. Usually, such an approach, by somewhat understating the estimate of the non-damaging level, provides the possibility of the guaranteed lower-level estimation of the probability of uptime under repeated actions of extreme loads.

The schematization method of the object's external extreme loads used to built reliability models implements a popular approach [9, 32, 36], which consists in replacing the continuous random loading process by a discrete sequence (flow) of stochastically independent random by size and identically distributed [55] influences $P_{H_1}, \ldots, P_{H_m}$ at random times $t_1, \ldots, t_m$. The corresponding scheme of the stationary discrete flow is shown in Fig. 1.1b. Considering that in practice, when the mobile equipment is used, extreme loads occur rarely, and the danger of sudden quasi-static destruction is mainly determined by the maximum value of equivalent stresses that arise during short-term loads; this way of schematizing the extreme load can be considered reasonable. Its use when reliability models are designed greatly simplifies the mathematical apparatus and in a number of cases allows you to obtain results in an analytical form important for engineering calculations.

Another approach to solving the reliability prediction problems during overload failures, based on the classical theory of emissions generated by a continuous stationary Gaussian random process [31], leads to well-known expressions for the probability of no-failure operation, given, for example, in [37, 38]. However, practical application of these formulas has been limited because of their proximity and the need to specify the parameters of the loading process, which are not specific to the engineering practice. In addition, when schematizing the process of extreme loads with a discrete sequence of random variables, the concept of the presence of the main (stationary) and additional extreme component of the loading process is used; they are assumed to be independent of each other. Such approach for many elements of transport, roadbuilding, agricultural and other mobile machines allows adequately reflecting real operational load conditions.

The general scheme of the sudden (overload) failure of the element used in the future is that for any number of extreme loads, the failure of an element occurs only in the case of at least a one-time excess of the extreme load $P_{\mathrm{H}i}$ of a fixed random level P_{Π} of the limit load of the element.

The scheme of sudden failures above corresponds to the mechanical model of the quasistatic deformation and fracture. It is assumed that the magnitude of the limit load of each element is completely determined by its initial quality and fixed in time, i.e. does not take into account the effect of aging. It is also supposed that the loss of the limit load is of a quasi-static nature and does not depend on the rate of deformation and other dynamic effects. The model considered excludes the effect on the limit load of the deformed element occurred earlier under its previous extreme loads. The above assumptions and limitations must be taken into account in the practical use of the following methods for predicting mechanical reliability and solving engineering problems to ensure it.

1.2.1 Reliability Models of Elements

Random load and constant deterministic limit load. Predicting reliability in sudden mechanical failures is associated with designing models that allow us to estimate the probability of no-failure operation, as the probability of an increase in the value of the repeated extreme load P_{H} of the magnitude of the P_{Π} element's limit load. Let us first consider a simplified version, when the time-constant limit load has such a small random dispersion that they can be neglected in practice. Consequently, for all instances of the elements considered, the limit load is the same and preserved in time. Then we assume that the nonrandom (deterministic) and time-constant limiting level of the P_o limit load is given. Considering that $P_o = \widetilde{P}_o$ (see Fig. 1.1a), such an assumption can be applied, i.e. using the lower limit of the possible random dispersal of the limit load as a limit level. Obviously, by doing this, we will obviously overestimate the projected risk of a sudden failure. In this case, the ratio of the limiting level P_o to the average value of random extreme loads $\overline{P}_{\mathrm{H}}$ should be taken as a safety factor K, i.e. $K = {}^{P_o}\!/\!_{\overline{P}_{\mathrm{H}}}$.

According to Chebyshev's inequality, known from the probability theory [10], for any distribution law of a positive random load and a constant limit load, it is possible to estimate the upper limit for the probability of failure under the first (single) extreme load:

$$\widetilde{Q_1} = \Pr(P_H \geq P_0) \leq \frac{\overline{P_H}}{P_0} = \frac{1}{K}. \tag{1.1}$$

Then, the lower limit $\widetilde{R}_1 = 1 - \widetilde{Q}_1$ for the probability of no-failure operation at the first load R_1 can be determined from the expression:

$$R_1 \geq \widetilde{R}_1 = 1 - \frac{1}{K}. \tag{1.2}$$

(1.2) concludes that if the information on the form and distribution parameters of extreme loads is absent, the guaranteed estimate $\widetilde{R}_1$, obtained under the assumption of arbitrary randomness of the load, is practically of little use to provide a sufficiently high standard level of no-failure operation, usually required for machine elements [6]: $[R] = 0.9 \div 0.999$. So, for example, if you set $\widetilde{R}_1 = 0.9$, then the corresponding value of the safety factor should be $K = 10$. The practice of designing machine-building products shows that it is not advisable to realize such durability in mobile machines. It should also be noted that an estimate of the type (1.2) can be justified only during a single extreme load, and during a repeated load, the specified reliability is not provided.

There is a rational approach, which makes it possible for the probabilistic justification of the safety coefficients and obtaining practically realizable recommendations. This approach consists in refusing to use the assumptions about their randomness when loads are schematized, but in specifying a type and parameters of the assumed law of distribution of random extreme effects. In this case, it is expedient to use the distribution functions of continuous positive random variables with a unimodal (single-vertex) density and an infinite upper bound for random scattering. At the design stage, the main and most informative parameter that determines the level of random dispersion of the load or limit load for machine-building reliability calculations [39] is the dimensionless variation coefficient. Therefore, it is preferable to apply two-parameter distribution laws. In the future when models are designed, the following laws will be widely used: the Weibull distribution law in the form [40], as well as the logarithmic logistic distribution [11] and the Frechet distribution [12]. The application of normal [39], log-normal [30], and double exponential [13] distributions is traditional.

Types and characteristics of some applicable distribution laws. Practicality in the engineering analysis of reliability of the distribution laws considered below is largely determined by the fact that the level of dispersion of a random variable relative to the mean for these distributions is uniquely estimated by only one parameter, which is called the shape parameter.

The logarithmic logistic distribution function *F*(*P*) and the distribution density *f*(*P*) look like:

$$F(P) = \frac{(P/C)^{\nu}}{1 + (P/C)^{\nu}}; \quad f(P) = \frac{V(P/C)^{\nu-1}}{C[1 + (P/C)^{\nu}]^2}, \tag{1.3}$$

where C is a scale parameter that is the median of the distribution; ν—shape parameter that uniquely determines the variation coefficient.

The average value of a random variable P at $\nu > 1$ is given by

$$\overline{P} = C \cdot \Gamma(1 + 1/\nu) \cdot \Gamma(1 - 1/\nu) = \frac{C \cdot \pi}{V \cdot \sin\frac{\pi}{\nu}}. \tag{1.4}$$

The variation coefficient, which is the ratio of the mean square deviation (the root of the dispersion) to the mean at $\nu > 2$ is determined from the expressions

$$V = \sqrt{\frac{\Gamma(1 + 2/\nu) \cdot \Gamma(1 - 2/\nu)}{\Gamma^2(1 + 1/\nu) \cdot \Gamma^2(1 - 2/\nu)} - 1} = \sqrt{\frac{\nu}{\pi} tg \frac{\pi}{\nu} - 1}. \tag{1.5}$$

To estimate the value of the shape parameter ν to the variation coefficient V at $V \leq 0.5$, we can use the approximate formula:

$$\nu \approx \pi \left[(1 - 0.3V^{11/3}) \sqrt{3.75\sqrt{\frac{1}{9} + \frac{8}{15}V^2} - 1.25} \right]^{-1} \tag{1.6}$$

At $\nu > 1$, the modal (most probable) value of P is determined from the expression:

$$\widehat{P} = C\left(\frac{1 - 1/\nu}{1 + 1/\nu}\right)^{1/\nu}. \tag{1.7}$$

In the case of the Frechet law, one of the limiting laws for the distribution of maxima, the distribution function and density look like:

$$F(P) = \exp\left[-\left(\frac{h}{P}\right)^{\rho}\right]; \quad f(P) = \frac{\rho}{P}\left(\frac{h}{P}\right)^{\rho} \exp\left[-\left(\frac{h}{P}\right)^{\rho}\right], \tag{1.8}$$

where h—a scale parameter; ρ—a shape parameter, which uniquely depends on the variation coefficient.

The average value and the variation coefficient of the Frechet distribution are calculated by the formulas:

$$\overline{P} = h\Gamma\left(1 - {}^{1}\!/_{\rho}\right) = \frac{h\pi}{\rho \sin\left({}^{\pi}\!/_{\rho}\right) \cdot \Gamma\left(1 + {}^{1}\!/_{\rho}\right)}, \quad \text{at } \rho > 1. \tag{1.9}$$

$$V = \sqrt{\frac{\Gamma\left(1 - {}^{2}\!/_{\rho}\right)}{\Gamma^2\left(1 - {}^{1}\!/_{\rho}\right)} - 1} = \quad \text{at } \rho > 2. \tag{1.10}$$

The distribution median (1.8) is given by

$$P_{0.5} = {}^{h}\!/_{(\ln 2)^{1/\rho}} \tag{1.11}$$

A modal value is given by the formula

$$\widehat{P} = h\left(\frac{\rho}{1+\rho}\right)^{1/\rho}. \tag{1.12}$$

Therefore, the scale parameter is in the range: $\widehat{P} < h < P_{0.5}$ and with an increase in the shape parameter ρ this interval narrows.

In a two-parameter Weibull distribution, the function and the distribution density are given by the formulas:

$$F(P) = 1 - \exp\left[-\left(\frac{P}{a}\right)^b\right]; \quad f(P) = \frac{b}{a}\left(\frac{P}{a}\right)^{b-1} \exp\left[-\left(\frac{P}{a}\right)^b\right], \tag{1.13}$$

where a—a scale parameter; b—a shape parameter.

Then the average value and variation coefficient are determined from the expressions

$$\overline{P} = a\Gamma\left(1 + {}^{1}\!/_{b}\right); \tag{1.14}$$

$$V = \sqrt{\frac{\Gamma\left(1 + {}^{2}\!/_{b}\right)}{\Gamma\left(1 + {}^{1}\!/_{b}\right)} - 1}. \tag{1.15}$$

The estimation of the shape parameter b by the variation coefficient in the case when $0.08 \leq V \leq 0.5$, can be approximately fulfilled [41] by the formula:

$$b \approx \frac{1.25}{V} - 0.4. \tag{1.16}$$

In a wider range of values of the variation coefficient at $0.08 \leq V \leq 1$, the approximate formula proposed in [14] can be used:

$$b \approx \frac{1.126}{V} + \frac{0.011}{V^2} - 0.137. \tag{1.17}$$

For the same variation coefficients at $0.06 \leq V \leq 0.4$ between the shape parameters of the laws of Frechet and Weibull, an approximate relation is valid: $\rho \approx b + 1.47$.

The median of the Weibull law is determined from expression

$$P_{0.5} = a(\ln 2)^{1/b}. \tag{1.18}$$

The modal value for b > 1 is determined by the formula

$$\widehat{P} = a\left(\frac{b-1}{b}\right)^{1/b}. \tag{1.19}$$

The distributions of Frechet and Weibull can be considered interrelated. If a random variable x has Frechet distribution, then the reciprocal $y = 1/x$ has Weibull distribution with the same shape parameter $b = \rho$, but with the scale parameter $a = 1/h$.

Table 1.1 gives some reference data on the shape parameters of the laws of Weibull, Frechet, and the log-logistic distribution, depending on the variation coefficient.

Double exponential distribution of the maxima has a function and density as:

$$F(P) = \exp\left\{-\exp\left(-\frac{P-\omega}{\beta}\right)\right\}; \tag{1.20}$$

$$f(P) = \frac{1}{\beta}\exp\left(-\frac{P-\omega}{\beta}\right) \cdot F(P), \tag{1.21}$$

where ω—a parameter that matches with a modal value.

Table 1.1 Value of shape parameters for two-parameter distributions

Kind of distribution	Shape parameter	Variation coefficient V								
		0.06	0.08	0.10	0.12	0.20	0.30	0.35	0.40	1.00
Weibull	b	20.68	15.35	12.15	10.03	5.80	3.71	3.13	2.70	1.00
Frechet	ρ	22.14	16.81	13.62	11.50	7.26	5.18	4.60	4.17	2.53
logarithmic logistic	ν	30.3	22.76	18.25	15.25	9.28	6.36	5.55	4.95	2.70

The average value is given by the formula:

$$\overline{P} = \omega + \beta C_o, \tag{1.22}$$

where $C_o = 0.5772156\ldots$—Euler constant.

The variation coefficient is determined from the formula:

$$V = \frac{\pi\beta}{\sqrt{6}(\omega + \beta C_o)}. \tag{1.23}$$

1.2.2 Estimated Reliability Function

If the load distribution function for any extreme load $F(P_H)$ is given and does not depend on the operating time and the number of extreme loads m, the reliability function R_1 with a single (first) load of the element can be determined as the probability of $P_H < P_o$ or

$$R_1 = F(P_o). \tag{1.24}$$

In a general case, at m repetitive extreme loads, the reliability function is determined by the probability that the constrained maximum of the m-multiple load $P_{max}(m) = \max(P_{H_1}, \ldots, P_{Hm})$ does not exceed the constant limit level P_o of the limit load of the element. The theory of extreme values of identically distributed independent random variables [13, 55] proves that the distribution functions of the constrained maximum $F(P_{max}(m))$ and loads $F(P_H)$ must be related by the dependence.

$$F(P_{max}(m)) = F^m(P_H). \tag{1.25}$$

Then the reliability function of an element when it is subjected to m-repeated extreme loads is given by the following expression:

$$R_m = F^m(P_o) = R_1^m = (1 - Q_1)^m, \tag{1.26}$$

where Q_1 is a probability of failure at the first load.

It follows from (1.26) that the number of extreme loads m to a sudden mechanical failure at a deterministic constant limiting level of the limit load of an element has a discrete geometric distribution [15, 48]. The probability function of failure at m loading or the discrete density of this distribution:

$$Q_m = R_{m-1} - R_m = Q_1(1 - Q_1)^{m-1}. \tag{1.27}$$

The average number of extreme loads to failure:

$$M = \frac{1}{Q_1}. \tag{1.28}$$

In this case, the discrete analogue of the intensity of sudden failures (the risk function) is defined by the expression: $\lambda_m = \frac{Q_m}{R_{m-1}} = Q_1$ and does not depend on m.

On the assumption of the fact that at a deterministic (nonrandom) limiting level of the limit load of an element, the number of loads prior to a sudden failure has geometric distribution. It is possible, using various laws of distribution of a random load, to obtain analytical expressions for predicting the indicators of mechanical reliability. Let us show the application of such an approach using the example of the use of the two-parameter Weibull law. Let the load distribution function look like

$$F(P_{\text{H}}) = 1 - \exp\left[-\left(\frac{P_{\text{H}}}{a_{\text{H}}}\right)^{b_{\text{H}}}\right]. \tag{1.29}$$

Then the reliability function at the first extreme load in accordance with (1.24) is determined from the expression:

$$R_1 = 1 - \exp\left[-\left(\frac{P_{\text{o}}}{a_{\text{H}}}\right)^{b_{\text{H}}}\right]. \tag{1.30}$$

Taking into account (1.14), the scale parameter a_{H} is expressed using the mean value of the extreme load $\overline{P}_{\text{H}}$ by the formula:

$$a_{\text{H}} = \frac{\overline{P}_{\text{H}}}{\Gamma\left(1 + {}^{1}\!/_{b_{\text{H}}}\right)}. \tag{1.31}$$

After substituting (1.31) into (1.30), taking into account that the safety factor is $K = \frac{P_o}{\overline{P}_{\text{H}}}$, we obtain an expression for the reliability function at the first load depending on the safety factor:

$$R_1(K) = 1 - \exp\left\{-\left[K\Gamma\left(1 + {}^{1}\!/_{b_{\text{H}}}\right)\right]^{b_{\text{H}}}\right\}. \tag{1.32}$$

Accordingly, the probability of failure at the first load Q_1 is determined from the expression:

$$Q_1(K) = \exp\left\{-\left[K\Gamma\left(1 + {}^{1}\!/_{b_{\text{H}}}\right)\right]^{b_{\text{H}}}\right\}. \tag{1.33}$$

Taking into account (1.28) from (1.33), we find that the average number of loads before a sudden failure is determined by formula:

$$M = \exp\left\{\left[K\Gamma\left(1 + {}^{1}\!/_{b_{\text{H}}}\right)\right]^{b_{\text{H}}}\right\}. \tag{1.34}$$

Taking into account (1.26) and using (1.32), we can obtain an equation for determining the safety factor $K_{\gamma}(m)$, which will provide the given reliability function γ for the known number of m extreme loads of the element:

$$\exp\left\{-\left[K_{r}(m)\cdot\Gamma\left(1 + {}^{1}\!/_{b_{\text{H}}}\right)\right]^{b_{\text{H}}}\right\} = 1 - \gamma^{1/m}. \tag{1.35}$$

The analytical solution of Eq. (1.35) is given in Table 1.2. This table also contains expressions for predicting the reliability function $R_{1}(K)$, the average number of failures M and probabilistically justified safety factors $K_{\gamma}(m)$ for different laws of distribution of extreme loads. In the cases of normal and log-normal distributions, the following notation is used in Table 1.2: $F_{o}(*)$—normalized distribution function [40]; V_{H}—a variation coefficient of extreme loads; $U_{\gamma}(m)$—a quantile of the normal distribution corresponding to the probability $\gamma^{1/m}$. To calculate $U_{\gamma}(m)$ at $\gamma \geq 0.5$, one can use [37] the approximate formula:

$$U_{\gamma} = \sqrt{\pi}\left\{\left(1 + \frac{2\gamma - 1}{181}\right)\left[\frac{1}{41}\left(\ln + {}^{1}\!/_{1-(2\gamma-1)^{4}}\right)^{3/2} - \ln 2\sqrt{\gamma(1-\gamma)}\right]\right\}^{1/2}. \tag{1.36}$$

Probabilistic justification of safety factors. Table 1.3 gives values of the safety factor $K_{\gamma}(m)$ for a number of variation coefficients of loads and the expected number of extreme loads, which ensure the reliability function $\gamma = 0.99$. The calculations were carried out using the expressions given in Table 1.2 for all the laws of load distribution listed in it. The data of Table 1.3 indicate that if the variation coefficient of the extreme load does not exceed 0.2 and the number of extreme loads is not more than 50, then for all variants of load distributions used to provide the above reliability function level of an element, it is sufficient to have approximately a threefold safety factor. An increase of the load variation coefficient requires higher safety factors.

It should be noted that the most probabilistically justified values of the safety factor were obtained using the Frechet law as the distribution of the extreme load. Given the significant influence of the load distribution type on the value of $K_{\gamma}(m)$, it is advisable to designate as much information as possible when designing. In critical cases when the distribution type of extreme loads is uncertain, the Fréchet distribution should be used.

Table 1.2 Dependencies for predicting mechanical reliability

Type of load distribution	Dependencies for prediction
1	2
Weibull	$R_1(K) = 1 - \exp\left\{-\left[K\Gamma\left(1 + 1/b_H\right)\right]^{b_H}\right\}$ $M = \exp\left\{\left[K\Gamma\left(1 + 1/b_H\right)\right]^{b_H}\right\}$ $K_\gamma(m) = \frac{\left(\ln\frac{1}{1-\gamma^{1/m}}\right)^{1/b_H}}{\Gamma\left(1+1/b_H\right)}$
Normal	$R_1(K) = F_o\left(\frac{K-1}{V_H}\right)$ $M = \left\{1 - F_o\left(\frac{K-1}{V_H}\right)\right\}^{-1}$ $K_\gamma(m) = 1 + U_\gamma(m)V_H$
Lognormal	$R_1(K) = F_0\left(\frac{\ln\left(K\sqrt{1+V_H^2}\right)}{\sqrt{\ln\left(1+V_H^2\right)}}\right)$ $M = \left\{1 - F_0\left(\frac{\ln\left(K\sqrt{1+V_H^2}\right)}{\sqrt{\ln\left(1+V_H^2\right)}}\right)\right\}^{-1}$ $K_\gamma(m) = \frac{1}{1+V_H^2}\exp\left\{U_\gamma(m)\sqrt{\ln\left(1+V_H^2\right)}\right\}$
1	2
Logarithmic logistic	$R_1(K) = \frac{K^{\nu_H}\left(\frac{\pi}{\nu_H \sin\frac{\pi}{\nu_H}}\right)^{\nu_H}}{1+K^{\nu_H}\left(\frac{\pi}{\nu_H \sin\frac{\pi}{\nu_H}}\right)^{\nu_H}}$ $M = 1 + K^{\nu_H}\left(\frac{\pi}{\nu_H \sin\frac{\pi}{\nu_H}}\right)^{\nu_H}$ $K_\gamma(m) = \frac{\gamma^{1/m\,\nu_H}\nu_H \sin\frac{\pi}{\nu_H}}{\pi\left(1-\gamma^{1/m}\right)^{1/\nu_H}}$
Double exponential	$R_1(K) = \exp\left\{-\exp\left[-\left(\frac{\pi(K-1)}{V_H\sqrt{6}} + C_o\right)\right]\right\}$ $C_o = 0,5772156\ldots$ $M = \left[1 - \exp\left\{-\exp\left(-\left(\frac{\pi(K-1)}{V_H\sqrt{6}} + C_o\right)\right)\right\}\right]^{-1}$ $K_\gamma(m) = 1 + V_H\left\{\ln\left[\left(\ln\frac{1}{\gamma^{1/m}}\right)^{-\frac{\sqrt{6}}{\pi}}\right] - \frac{C_o\sqrt{6}}{\pi}\right\}$
1	2
Frechet	$R_1(K) = \exp\left\{-\left[\frac{\rho_H \sin\left(\frac{\pi}{\rho_H}\right)\cdot\Gamma\left(1+1/\rho_H\right)}{\pi K}\right]^{\rho_H}\right\}$ $M = \left[1 - \exp\left\{-\left[\frac{\rho_H \sin\left(\frac{\pi}{\rho_H}\right)\cdot\Gamma\left(1+1/\rho_H\right)}{\pi K}\right]^{\rho_H}\right\}\right]^{-1}$ $K_\gamma(m) = \frac{\rho_H m^{1/\rho_H} \sin\left(\frac{\pi}{\rho_H}\right)\cdot\Gamma\left(1+1/\rho_H\right)}{\pi\left(\ln\frac{1}{\gamma}\right)^{1/\rho_H}}$

Table 1.3 Probabilistically justified safety factors at $\gamma = 0.99$

Kind of distribution load	Number of loadings m	Variation coefficient of loads 0.06	0.10	0.20	0.30
Weibull	1	1.105	1.183	1.406	1.672
	10	1.127	1.223	1.507	1.865
	50	1.139	1.244	1.562	1.974
Normal	1	1.140	1.233	1.466	1.698
	10	1.185	1.309	1.618	1.926
	50	1.213	1.354	1.708	2.062
Lognormal	1	1.146	1.249	1.525	1.817
	10	1.199	1.347	1.772	2.271
	50	1.232	1.410	1.938	2.594
Logarithmic logistic	1	1.162	1.280	1.610	1.977
	10	1.254	1.453	2.064	2.841
	50	1.322	1.586	2.455	3.660
Double exponential	1	1.188	1.314	1.627	1.941
	10	1.296	1.493	1.986	2.480
	50	1.371	1.619	2.237	2.856
Frechet	1	1.197	1.338	1.712	2.102
	10	1.328	1.584	2.350	2.887
	50	1.428	1.783	2.934	4.472

1.3 Dependences of Reliability Indices on Operating Time

The transition in the measurement of the volume of the produced resource from the number of extreme loads m to the operating time t can be carried out by accepting certain assumptions about the type of a random flow of instants of discrete loads (see Fig. 1.1b). In many cases, assuming the independence of random operating times before loading, $t_1, t_2, \ldots, t_m$ the assumption of a stationary Poisson flow is acceptable, where the density of the distribution of the number of loads i per the operating time t has the form

$$P_i(t) = \frac{e^{-\frac{t}{T_O}}}{i!} \cdot \left(\frac{t}{T_o}\right)^i, \tag{1.37}$$

where T_o an average period between random loads.

Then, based on the well-known formula for full probability, we have an expression that determines the reliability function as a function of the safety factor and operating time:

$$R(K, t) = \sum_{i=0}^{\infty} P_i(t) R_i(K), \tag{1.38}$$

where, in accordance with (1.26), $R_i(K) = R_1^i(K)$—a conditional reliability function for a random number i of loads.

It follows from (1.37) and (1.38) that the operating time to failure as an exponential distribution in this case, and the reliability function is determined by the expression:

$$R(K,t) = \exp\left(-\frac{t}{T_o}(1 - R_1(K))\right). \tag{1.39}$$

In this case, the mean time to failure is determined by the formula

$$T = \frac{T_o}{1 - R_1(K)} = MT_o, \tag{1.40}$$

and the failure rate is independent of the operating time constant:

$$\lambda = \frac{1 - R_1(K)}{T_o} = (1 - R_1(K))\omega_o, \tag{1.41}$$

where $\omega_o = 1/T_o$ is a constant rate of extreme loads.

The expression (1.41) indicates that for the constant nonrandom limit load and a stationary Poisson load flow, the stream of sudden failures in the set of elements is also a stationary Poisson flow, but with a reduced rate.

The gamma-percentile operating time to failure can be determined by the formula

$$t_\gamma(K) = \frac{T_o \ln 1/\gamma}{1 - R_1(K)} = T \ln 1/\gamma. \tag{1.42}$$

It follows from (1.42) that the relative value of the gamma-percentile operating time to failure $\tau_\gamma(K) = t_\gamma(K)/T_o$ depends on the safety factor and the type of distribution of extreme loads, which determines the reliability function at the first extreme load $R_1(K)$ (see Table 1.2). It can be concluded from the analysis of the data given in Table 1.3 that the predicted value $\tau_\gamma(K)$ will be the smallest, if one uses $R_1(K)$ the Frechet distribution from the considered laws. This gives some guarantees against overstating $\tau_\gamma(K)$ when a type of the load distribution is uncertain. This approach is used in calculating the values $\tau_\gamma(K)$ given in Table 1P of the appendixes.

When predicting gamma-percentile time to the failure $t_\gamma(K)$, the tabulated values $\tau_\gamma(K)$ should be multiplied by the expected value of the mean period T_o between the extreme loads. For example, if when designing it can be specified $T_o = 300$ h, and the variation coefficient of extreme loads $V_H = 0.1$, then for the value of the safety factor K $= 1.8$ for $\gamma = 0.99$, it follows from the Table 1P of the appendixes that $\tau_\gamma = 57.14$. Then the predicted gamma-percentile operating time to failure is: $t_\gamma = 57.14 \cdot 300 = 17,142$ hours. Therefore, in practice during such a lifetime, no more than 1% of the designed elements may have sudden failures on average. If it is necessary to consider the possibility of a significant dispersion of values of extreme loads, and set the variation coefficient $V_{\text{H}} = 0.2$, then when increasing the safety factor to $K = 2.4$ and setting $\gamma = 0.95$, we receive a table value $\tau_\gamma = 59.44$ and respectively a gamma-percentile time $t_\gamma = 59.44 \cdot 300 = 17,832$ h. In this case, it is necessary to take into account that during the lifetime, a predicted risk of failures increases to 5%, which is the cause of an increase in load variation coefficient.

In the general case of the Poisson flow of extreme loads, the distribution density of loads i for operating time t is determined by the following formula:

$$P_i(t) = \frac{e^{-\overline{m}(t)} \cdot (\overline{m}(t))^i}{i!}, \tag{1.43}$$

where $\overline{m}(t)$ an average total number of extreme loads per operating time t.

For a stationary flow $\overline{m}(t) = \omega_o t$ and, consequently, the total number of loads is a linear function of the operating time. In the general case, the relationship between $\overline{m}(t)$ and a variable intensity of the loads ωt is as follows

$$\overline{m}(t) = \int_0^t \omega(t)dt. \tag{1.44}$$

If the total average of the nonstationary Poisson flow $\overline{m}(t)$ is a monotonically increasing power function of the operating time $\overline{m}(t) = \omega_o t^\nu$, then the variable intensity of loads has the following form:

$$\omega(t) = \frac{d\overline{m}(t)}{dt} = \omega_o \nu t^{\nu-1}. \tag{1.45}$$

Using (1.43) and (1.38), for the reliability function we obtain an expression as follows:

$$R(K,t) = \exp\{-(1 - R_1(K))\omega_o t^\nu\}. \tag{1.46}$$

The expression (1.46) indicates that in this case the operating time to failure has a Weibull distribution with a shape parameter ν. A random Poisson flow of extreme loads with an intensity of the form (1.45) is called [16, 17] the Weibull flow. The mean time to failure is determined using the expression:

$$T = \frac{\Gamma(1 + 1/\nu)}{[\omega_o(1 - R_1(K))]^{1/\nu}}. \tag{1.47}$$

The gamma-percentile operating time to failure is determined using the expression:

$$t_\gamma = \left\{ \frac{\ln 1/\gamma}{\omega_o(1 - R_1(K))} \right\}^{1/\nu}. \tag{1.48}$$

It follows from (1.46) that the dependence of the failure rate on the operating time has the following form:

$$\lambda(t) = \omega_o \nu (1 - R_1(K)) t^{\nu - 1}. \tag{1.49}$$

At $\nu > 1$, the failure rate as well as the loading intensity will be an increasing function of the operating time, and at $\nu < 1$ it will monotonically decrease. Consequently, with the geometric distribution of the number of extreme loads prior to a sudden failure and the Poisson load flow, the character of the monotonic change in the failure rate as a function of the operating time is completely determined by the nature of the change in the corresponding intensity of the extreme loads. The failure rate is lowered due to the multiplier $1 - R_1(K)$ that depends on the safety factor, a type of distribution, and the variation coefficient of the load (see Table 1.2). An increase of the safety factor leads to a decrease in the failure rate in comparison with the intensity of the load flow.

Chapter 2
Reliability of Elements with Random Limit Load

2.1 Reliability Models Under Multiple Loadings

In general, building models to predict the reliability of elements in sudden failures should take into account both randomness of extreme loads and the random nature of the limit load [29, 30, 32, 39, 41, 42]. Practice and test results show that the limit load of mechanical elements has a random dispersion with a variation coefficient from 0.05 to 0.15. Therefore, when predicting the reliability of an element as for the reliability function in the case of a single (first) extreme load, the probability is assumed that, where $P_{\Pi} > P_{\mathrm{H}}$, P_{Π} and P_{H} are random values of the limit load and loads.

If we consider the values of the limit load and the load independent, then under single (first) loading and normal distribution P_{Π} and P_{H}, it is convenient to use the condition of reliability in the form

$$z = P_{\Pi} - P_{\mathrm{H}} > 0. \tag{2.1}$$

Considering (2.1), it is possible to obtain [30, 43] the following expression for the reliability function at the first load of the element

$$R_1 = F_{\mathrm{o}}\left(\frac{\overline{K} - 1}{\sqrt{V_{\Pi}^2 \cdot \overline{K}^2 + V_{\mathrm{H}}^2}}\right), \tag{2.2}$$

where $\overline{K} = \frac{\overline{P}_{\Pi}}{\overline{P}_{\mathrm{H}}}$ the safety factor calculated from the average values of the limit load and load;

$F_{\mathrm{o}}(*)$—normalized distribution function;

V_{Π}—a variation coefficient of the limit load.

Setting the probability value $R_1 = \gamma$, we obtain the well-known [30] expression to determine the probabilisticaly justified safety factor from (2.2).

O. Grynchenko, O. Alfyorov, *Mechanical Reliability*,
https://doi.org/10.1007/978-3-030-41564-8_2

$$\overline{K}(\gamma) = \frac{1 + U_{\gamma}\sqrt{V_{\Pi}^{2} + V_{H}^{2} - (U_{r} \cdot V_{\Pi} \cdot V_{H})^{2}}}{1 - U_{\gamma}^{2} \cdot V_{\Pi}^{2}}. \tag{2.3}$$

where U_{γ} a quantile of the normal distribution corresponding to the probability γ (see formula (1.36)).

In the case of a logarithmically normal distribution of the limit load and load, it is more convenient to use one more variant of the condition of failure-free operation at the first load.

$$z = {}^{P_{\Pi}}/_{P_{H}} > 1, \tag{2.4}$$

which after taking the logarithm takes a similar (2.1) form

$$\ln z = \ln P_{\Pi} - \ln P_{H} > 0. \tag{2.5}$$

Then

$$R_1 = F_o\left(\frac{\ln\overline{K} - \ln\sqrt{\frac{1+V_{\Pi}^2}{1+V_{H}^2}}}{\sqrt{\ln\left[\left(1+V_{\Pi}^2\right)\left(1+V_{H}^2\right)\right]}}\right), \tag{2.6}$$

and the formula for a probabilistically justified safety factor has the form [30]:

$$\overline{K}(\gamma) = \sqrt{\frac{1+V_{\Pi}^2}{1+V_{H}^2}}\exp\left(U_{\gamma}\sqrt{\ln\left[\left(1+V_{\Pi}^2\right)\left(1+V_{H}^2\right)\right]}\right). \tag{2.7}$$

When using different types of distributions of the random load values and the limit load of the elements to predict the reliability function at the first load, the following expressions [32, 44, 45] are widely used:

$$R_1 = \int_0^{\infty} F(P)g(P)dP, \tag{2.8}$$

$$R_1 = \int_0^{\infty} [1 - G(P)]f(P)dP = 1 - \int_0^{\infty} G(P)f(P)dP, \tag{2.9}$$

where *F*(*P*) and *f*(*P*)—a function and a density of the load distribution; *G*(*P*) and *g* (*P*)—a function and density of the limit load distribution.

Formulas (2.8) and (2.9) assume that the load and the limit load are independent random variables. The meaning of the formulas is equivalent and the choice between them is determined only by the ease of use for specific types of distributions.

Let us consider the most frequently encountered variant of the randomness of the quasi-static limit load of machinery elements and structures, which consists in the fact that the random limit load of each element is determined only by its initial quality and practically independent of the operating time or number of loads to which the element was subjected. In this case, the form and the distribution parameters of the limit load do not depend on the operating time. If loading occurs repeatedly and the limit load of the element is a random variable that does not change in time, then proceeding from the fact that the distribution function of the maximum of the discrete random sequence of loads $P_{H_1}, \ldots, P_{H_m}$ (see Fig. 1.1b) in accordance with (1.25) is expressed as $F^m(P)$. By analogy with (2.8), one can obtain [32] an expression for the reliability function of an element with m-fold loading:

$$R_m = \int_0^{\infty} F^m(P)g(P)dP. \tag{2.10}$$

Taking into account that the density distribution of the maximum of the random load sequence is as follows:

$$f(P_{\max}(m)) = \frac{dF^m(P)}{dP} = mF^{m-1}(P)f(P), \tag{2.11}$$

By analogy with (2.9) we obtain the equivalent expression (2.10) for the reliability function:

$$R_m = m\int_0^{\infty} [1 - G(P)]F^{m-1}(P)f(P)dP. \tag{2.12}$$

The reliability models (2.10) and (2.12) assume that for each element its random limit load is realized only once, remaining a fixed value in time in the process of loading this element. The random load acting on the element is implemented m-fold, forming a random discrete sequence of independent quantities (Fig. 1.1b). It is also assumed that the load distributions and the limit load do not depend on time (operating time).

Another less common option of the randomness of the limit load of an element is possible, when it is not fixed in time at random moments of extreme loads t_1, t_2, …, t_m, but takes on the changing values formed by the realization of an ergodic stationary random process. We can assume that such a value of the limit load is independent random variables with the distribution density $g(P)$. Then the reliability function at the first load R_1 should be determined by the formulas (2.8) or (2.9), and the probability R_m at them-fold loading can then be determined using the expression (1.26), i.e. in this case, as with a constant deterministic limit load $R_m = R_1^m$.

2.2 The Method of Transition to Unit Distributions

Practical application of models of the form (2.8), (2.9), (2.10), and (2.12) can be simplified using the transformation, called a transition to unit distributions [24]. The expressions (2.8) and (2.9) can be rewritten in the form

$$R_1 = \int_0^\infty F(P)dG(P), \tag{2.13}$$

$$R_1 = 1 - \int_0^\infty G(P)dF(P). \tag{2.14}$$

Obviously, for the transition to (2.13) from integration over a variable P to integration over a variable G, it is sufficient to eliminate a variable P in $F(P)$, expressing it with the help of a quantity G. We denote by the symbol $\psi(G)$ the function inverse to the distribution function of the limit load $G(P)$. This function determines the magnitude of the quantiles of the limit load distribution $P_G = \psi(G)$ that correspond to a probability G. We will call it the inverse distribution function of the limit load. The variable G varies in the unit interval: $0 \leq G \leq 1$. Therefore, being obtained by substituting for the load distribution function $F(P)$ instead of its argument P of the inverse bearing load distribution function $\psi(G)$, the complex function will be called the function of the unit load distribution:

$$F_1(G) = F(\psi(G)). \tag{2.15}$$

An essential feature of the distribution resulting from such a transformation of distribution is that it has not only a function, but the argument varies in the interval from zero to one. Therefore, the expression (2.13) for a reliability function of the element at the first load after the transition to the unit load distribution takes the following form:

$$R_1 = \int_0^1 F_1(G)dG. \tag{2.16}$$

The expression (2.14) can also be transformed in a similar way if we go over to the function of the unit distribution of the limit load:

$$G_1(F) = G(\varphi(F)), \tag{2.17}$$

where $\varphi(F)$ is an inverse of the load distribution function $F(P)$.

Then instead of (2.14) we get the equivalent expression:

$$R_1 = 1 - \int_0^1 G_1(F)dF. \tag{2.18}$$

The transition to unit distributions depending on which function of unit distributions—$F_1(G)$ or $G_1(F)$ is more convenient to use, also transforms the general expressions (2.10) and (2.12), which gives four versions of equivalent formulas for the reliability function of an element under m-fold loading:

$$R_m = \int_0^1 F_1^m(G)dG = \int_0^1 F^m g_1(F)dF; \tag{2.19}$$

$$R_m = 1 - m\int_0^1 GF_1^{m-1}(G)f_1(G)dG = 1 - m\int_0^1 G_1(F)F^{m-1}dF, \tag{2.20}$$

where $f_1(G) = \frac{dF_1(G)}{dG}$—a density of the unit load distribution;
$g_1(F) = \frac{dG_1(F)}{dF}$—a density of the unit distribution of the limit load.

For various types of load distributions and limit load, forecasting the probability of fail-safe operation can always be performed by numerical integration of expressions (2.19) or (2.20) on a finite (unit) interval [18]. In some particular cases, it is easier to obtain analytical results by the method of transition to unit distributions (2.15) and (2.17).

Reference data on the functions of single distributions $F_1(G)$ and $G_1(F)$, which correspond to some combinations of load distributions and a limit load, are given in the appendixes (see Table 2P). This table contains analytical expressions that depend not only on the single arguments G and F, but also on the safety factor determined by the average values of the limit load and load: $\overline{K} = \overline{P}_{\Pi}/\overline{P}_H$

2.3 Models Based on Weibull Distribution

Let us consider some examples of using the method of single distributions. If the Weibull law is used as the functions of load distribution and limit load:

$$F(P) = 1 - \exp\left[-\left(\frac{P}{a_{\mathrm{H}}}\right)^{b_{\mathrm{H}}}\right]; \quad G(P) = 1 - \exp\left[-\left(\frac{P}{a_{\Pi}}\right)^{b_{\Pi}}\right], \tag{2.21}$$

then the inverse load distribution function $\psi(G)$ will have the form

$$\psi(G) = a_{\Pi}\left(\ln\frac{1}{1-G}\right)^{1/b_{\Pi}}. \tag{2.22}$$

Then the function of the unit load distribution is determined from the expression:

$$F_1(G) = 1 - \exp\left\{-\left(\frac{a_{\Pi}}{a_{H}}\right)^{b_H}\left(\ln\frac{1}{1-G}\right)^{b_H/b_{\Pi}}\right\}. \tag{2.23}$$

In the particular case where the random variables and limit load similar to [37], and the coefficients of variation of the limit load and equal or close in magnitude: $b_H \approx b_{\Pi} = b$, then the safety factor by average $\overline{K} = \frac{a_{\Pi}}{a_H}$. Then the expression for the function of the unit load distribution becomes simpler and takes the form

$$F_1(G) = 1 - (1 - G)^{\overline{K}^b}. \tag{2.24}$$

Determining the reliability function at the first load from (2.16), we obtain that

$$R_1 = \int_0^1 \left[1 - (1 - G)^{\overline{K}^b}\right] dG = 1 - \int_0^1 (1 - G)^{\overline{K}^b} dG. \tag{2.25}$$

The integral in (2.25) is calculated by changing the variable $x = 1 - G$, which gives the formula:

$$R_1 = \frac{\overline{K}^b}{\overline{K}^b + 1}. \tag{2.26}$$

Assuming that the load and the limit load have a Weibull distribution (2.21), the inverse load distribution function $\varphi(F)$ takes the form

$$\varphi(F) = a_H\left(\ln\frac{1}{1-F}\right)^{1/b_H}, \tag{2.27}$$

and the function of the unit distribution of the limit load is determined from the expression:

$$G_1(F) = 1 - \exp\left\{-\left(\frac{a_H}{a_{\Pi}}\right)^{b_{\Pi}}\left(\ln\frac{1}{1-F}\right)^{b_{\Pi}/b_H}\right\}. \tag{2.28}$$

In the particular case at $b_H \approx b_{\Pi} = b$ when we get that

$$G_1(F) = 1 - (1 - F)^{1/\overline{K}^b}. \tag{2.29}$$

Substitution of (2.29) into the second version of formulas (2.20) gives an expression for the reliability function for m-fold loading

$$R_m = m\int_0^1 F^{m-1}(1-F)^{1/\overline{K}^b}\,dF. \tag{2.30}$$

The integral on the right-hand side of (2.30) is a beta function [46]. Consequently,

$$R_m = m\cdot B\left(m, 1+1/\overline{K}^b\right). \tag{2.31}$$

Expressing the beta function with the help of gamma functions, we obtain an analytical dependence [14] for the reliability function

$$R_m = \frac{\Gamma(m+1)\Gamma\left(1+1/\overline{K}^b\right)}{\Gamma\left(m+1+1/\overline{K}^b\right)}. \tag{2.32}$$

Taking into account that the following formulas are relevant for integers m:

$$\begin{aligned} &\Gamma(m+1) = m!;\\ &\Gamma\left(m+1+1/\overline{K}^b\right) = \Gamma\left(1+1/\overline{K}^b\right)\cdot\prod_{i=1}^{m}\left(i+1/\overline{K}^b\right), \end{aligned} \tag{2.33}$$

substituting them into (2.32), we obtain an expression for the reliability function under repeated loads, which is convenient for practical use in engineering calculations:

$$R_m = \prod_{i=1}^{m}\frac{i\overline{K}^b}{i\overline{K}^b+1}. \tag{2.34}$$

One important conclusion should be pointed out, which follows from the formula (2.34). It turns out that if random scattering of the limit load of elements is present, the number of extreme loads prior to a sudden failure does not obey the geometric distribution (1.26). Indeed, it follows from (2.26) and (2.34) that for all $m > 1$ the following inequality is obvious:

$$R_m = \prod_{i=1}^{m}\frac{i\overline{K}^b}{i\overline{K}^b+1} > \left(\frac{\overline{K}^b}{\overline{K}^b+1}\right)^m = R_1^m.$$

Using (2.34), we can obtain an expression for the conditional reliability function of an element that has already withstood without failure the first m_{Π} preliminary load:

$$R\left({}^{m}\!/_{m_{\Pi}}\right) = \frac{R_{m_{\Pi}+m}}{R_{m_{\Pi}}} = \prod_{i=m_{\Pi}+1}^{m_{\Pi}+m} \frac{i\overline{K}^{b}}{i\overline{K}^{b}+1}. \tag{2.35}$$

An important conclusion from (2.35) is that an increase in the number of preloadsm_{Π}, starting with the first at constant m, leads to an increase in the conditional reliability function $R\left({}^{m}\!/_{m_{\Pi}}\right)$. This can serve as a justification for the application of a series of preloads (power "break-in") to improve the reliability of elements. This manifests an essential distinctive property of elements that have random scattering of the limit load, which is a fixed random variable but independent of time for each element. In the case of deterministic limit load considered in Chap. 1 in accordance with (1.26), the conditional reliability function does not differ from the unconditional one:

$$R\left({}^{m}\!/_{m_{\Pi}}\right) = \frac{R_1^{m_{\Pi}+m}}{R_1^{m_{\Pi}}} = R_1^{m} = R_m. \tag{2.36}$$

Therefore, the power break-in cannot give an effect here. Logically, this is quite understandable, because if all elements of the limit load are the same, then the screening of the "weakest" is impossible.

Formula (2.34) is simple enough, but even for $m > 5$, its direct application in engineering calculations can cause difficulties. Therefore, it is advisable to convert it to a form more convenient for practical calculations for bigger m. After taking the logarithm of the expression (2.34), we obtain that

$$\ln R_m = -\sum_{i=1}^{m} \ln\left(1 + \frac{1}{i\overline{K}^{b}}\right). \tag{2.37}$$

Expanding the logarithmic functions on the right-hand side of (2.37) into power series and restricting to the third power in the expansions, we have

$$\ln R_m \cong -\Omega \cdot S_1(m) + \frac{\Omega^2}{2} S_2(m) - \frac{\Omega^3}{3} S_3(m), \tag{2.38}$$

where $\Omega = 1/\overline{K}^{b}$; $S_1(m) = \sum_{i=1}^{m} \frac{1}{i}$; $S_2(m) = \sum_{i=1}^{m} \frac{1}{i^2}$; $S_3(m) = \sum_{i=1}^{m} \frac{1}{i^3}$.

Consequently, the reliability function at any number of extreme loads m can be calculated from the formula

$$R_m = \exp\left\{-\Omega \cdot S_1(m) + \frac{\Omega^2}{2} S_2(m) - \frac{\Omega^3}{3} S_3(m)\right\}. \tag{2.39}$$

Assuming that $\Omega < 0.1$ formula (2.39) provides a fairly good approximation to the exact expression (2.34). The values of the sums $S_1(m)$, $S_2(m)$ and $S_3(m)$ are given in Table 2.1. It follows from (2.39) an approximate expression for a probabilistically justified safety factor that takes into account the random scattering of the limit load and provides a predetermined value of the reliability function γ due to a sudden failure at m extreme loads:

$$\overline{K}(m,\gamma) = \left\{ \frac{S_1(m)}{\ln 1/_{\gamma}} \right\}^{1/_b}. \tag{2.40}$$

For sufficiently large m (in practice, if $m > 5$), we can assume that$S_1(m) \cong C_o + \ln(m + 0.51)$, where $C_o = 0.57721...$—Euler constant. The sums $S_2(m)$ and $S_3(m)$ at $m \to \infty$ have finite limits: $\lim S_2 = \frac{\pi^2}{6}$and $\lim S_3 = 1.202057$. Therefore, as it can be seen from the data in Table 2.1, if the expected number of extreme loads is $m > 100$, then the design calculation of the reliability function can be carried out according to the formula

$$R_m = (m + 0.51)^{-\Omega} \cdot \exp\left(-C_o\Omega + \frac{\pi^2}{12} \cdot \Omega^2 - 0.4\Omega^3 \right). \tag{2.41}$$

Consequently, if $m > 5$, then instead of (2.40) we can use the approximate formula

$$\overline{K}(m,\gamma) = \left\{ \frac{C_o + \ln(m + 0.51)}{\ln 1/_{\gamma}} \right\}^{1/_b}. \tag{2.42}$$

In the general case, when the coefficients of variation of the load and limit load are different in magnitude and $b_{\text{н}} \neq b_{\text{п}}$, the numerical integration of expression (2.19) should be applied using (2.23). We represent (2.23) in the form:

$$F_1(G) = 1 - \exp\left(-\eta \cdot \left[\ln 1/_{(1-G)} \right]^{\beta} \right), \tag{2.43}$$

where $\eta = \left(\frac{a_{\text{п}}}{a_{\text{н}}} \right)^{b_{\text{н}}}$; $\beta = {b_{\text{н}}}/{b_{\text{п}}}$.

In contrast to the functions $G(P)$ and $F(P)$, which depend on the four distribution parameters $a_{\text{п}}$, $a_{\text{н}}$, $b_{\text{п}}$ and $b_{\text{н}}$, the function $F_1(G)$ of the unit load distribution depends only on two generalized dimensionless parameters η and β. This greatly simplifies the tabulation of the reliability function, calculated on the basis of (2.19) by numerical integration of the expression

$$R_m = \int_o^1 \left\{ 1 - \exp\left(-\eta \left[\ln 1/_{(1-G)} \right]^{\beta} \right) \right\}^m dG. \tag{2.44}$$

Table 2.1 Auxiliary coefficients for reliability calculations

m	$S_1(m)$	$S_2(m)$	$S_3(m)$
1	1.0	1.0	1.0
2	1.5	1.25	1.125
3	1.833333	1.361111	1.162037
4	2.083333	1.423611	1.177662
5	2.283333	1.463611	1.185662
6	2.45	1.491389	1.190292
7	2.592857	1.511797	1.193207
8	2.717857	1.527422	1.19516
9	2.828968	1.539768	1.196532
10	2.928968	1.549768	1.197532
12	3.103211	1.564977	1.198862
14	3.251562	1.575996	1.199682
16	3.380729	1.584347	1.200222
18	3.495108	1.590893	1.200597
20	3.59774	1.596163	1.200868
25	3.815958	1.605723	1.201288
30	3.994987	1.61215	1.20152
35	4.146781	1.616767	1.20166
40	4.278543	1.620244	1.201752
45	4.394948	1.622957	1.201815
50	4.499205	1.625133	1.201861
60	4.67987	1.628406	1.20192
70	4.832837	1.63075	1.201956
80	4.965479	1.632508	1.20198
90	5.082571	1.63388	1.201996
100	5.187378	1.63498	1.202007
∞	–	1.64493	1.202057

Setting the value of the reliability function of γ and by a numerical solution of the equation

$$\int_{o}^{1}\left\{1-\exp\left(-\eta_{\gamma}\left[\ln {}^{1}/_{(1-G)}\right]^{\beta}\right)\right\}^{m} dG=\gamma \tag{2.45}$$

for different values of the parameter β and the number of extreme loads m, we can define the corresponding value of the parameter $\eta_\gamma(m)$. This allows you to calculate the required value of the safety factor of the element, providing a given reliability function at m extreme loads, according to the formula

$$\overline{K}(\gamma,m)=\frac{\Gamma\left(1+{}^{1}/_{b_{\Pi}}\right)}{\Gamma\left(1+{}^{1}/_{b_{H}}\right)}\left[\eta_{\gamma}(m)\right]^{1/b_{H}}. \tag{2.46}$$

Table 2.2 Data on the dependence of the parameter $\eta_\gamma(m)$ on the number of loads

$\beta = {}^{b_H}/_{b_\Pi}$	$\eta_{1\gamma}$ $(m = 1)$	$\varphi_\gamma(m)$ at the number of loadings m:					
		10	100	500	1000	2000	5000
0	$\frac{6.9078}{9.2103}$	$\frac{1.3333}{1.2500}$	$\frac{1.6666}{1.5000}$	$\frac{1.8996}{1.6747}$	$\frac{1.9999}{1.7500}$	$\frac{2.1003}{1.8253}$	$\frac{2.2329}{1.9247}$
0.1	$\frac{7.8}{10.9}$	$\frac{1.3846}{1.3028}$	$\frac{1.8205}{1.6514}$	$\frac{2.141}{1.9266}$	$\frac{2.2949}{2.055}$	$\frac{2.4231}{2.1743}$	$\frac{2.6282}{2.3486}$
0.3	$\frac{15.9}{29.6}$	$\frac{1.7862}{1.8682}$	$\frac{2.7862}{2.9797}$	$\frac{3.5409}{3.7973}$	$\frac{3.8742}{4.1486}$	$\frac{4.2075}{4.5101}$	$\frac{4.6541}{4.9899}$
0.5	$\frac{45.2}{136.4}$	$\frac{2.2257}{2.3321}$	$\frac{3.7376}{3.9172}$	$\frac{4.8363}{5.0682}$	$\frac{5.3142}{5.5682}$	$\frac{5.7920}{6.0696}$	$\frac{6.3783}{6.6840}$
0.7	$\frac{148}{740.2}$	$\frac{2.811}{2.5875}$	$\frac{4.4669}{4.4774}$	$\frac{5.7561}{5.7697}$	$\frac{6.4041}{6.4195}$	$\frac{6.9899}{7.0069}$	$\frac{7.7655}{7.7842}$
0.8	$\frac{277}{1748.4}$	$\frac{2.7108}{2.7111}$	$\frac{4.7368}{4.7368}$	$\frac{6.1838}{6.1837}$	$\frac{6.8094}{6.8092}$	$\frac{7.4354}{7.4353}$	$\frac{8.2635}{8.2636}$
0.9	$\frac{523.9}{4165}$	$\frac{2.8267}{2.8257}$	$\frac{4.9109}{4.9088}$	$\frac{6.5077}{6.5048}$	$\frac{7.1689}{7.1658}$	$\frac{7.8305}{7.8271}$	$\frac{8.7055}{8.7017}$
1.0	$\frac{999}{9999}$	$\frac{2.9193}{2.9182}$	$\frac{5.1193}{5.1171}$	$\frac{6.7236}{6.7205}$	$\frac{7.4277}{7.4244}$	$\frac{8.1391}{8.1353}$	$\frac{9.0883}{9.0840}$
1.2	$\frac{3692.9}{58582}$	$\frac{3.0824}{3.0813}$	$\frac{5.4897}{5.4877}$	$\frac{7.2596}{7.2568}$	$\frac{8.0384}{8.0353}$	$\frac{8.8242}{8.8207}$	$\frac{9.8708}{9.8668}$

Table 2.2 gives some results of the numerical solution of Eq. (2.45), which allows us to determine the value of the parameter $\eta_\gamma(m)$ as $\eta_\gamma(m) = \eta_{1\gamma} \cdot \varphi_\gamma(m)$ for the values of $\gamma = 0.999$ (numbers above the bar) and $\gamma = 0.9999$ (numbers below the bar).

So, for example, if $b_\Pi = 10$ (a variation coefficient of the limit load $V_\Pi \approx 0.12$), and $b_H = 5$ (variation coefficient of the load $V_H \approx 0.23$), and accordingly, $\beta = 0.5$, then for $\gamma = 0.999$ and $m = 1000$, proceeding from the table values $\eta_{1\gamma} = 45.2$ and $\varphi\gamma\ (1000) = 5.3142$, we get that $\eta_\gamma(1000) = 45.2 \cdot 5.3142 = 240.2$. The safety factor determined by formula (2.46), which ensures a reliability function of at least 0.999 with the number of loads $m = 1000$, in this case it is

$$\overline{K}(0.999; 1000) = \frac{\Gamma(1.1)}{\Gamma(1.2)} \cdot 240.2^{0.2} = 3.1.$$

It is also possible to obtain approximate analytical dependencies in the general case of the distribution of the limit load and load according to the Weibull law with different shape parameters, i.e. when the limit load and load are not similar to random variables and have different coefficients of variation.

Integration into (2.44) can be performed using the expansion of the integrands in a series. This allows us to obtain practically useful analytical results. The substitution of the variable $z = \eta[\ln {}^1/_{(1-G)}]^\beta$ and $m = 1$ leads the integral (2.44) to the form

$$R_1 = 1 - \frac{1}{\eta\beta} \int_0^\infty e^{-z} e^{-\left(\frac{z}{\eta}\right)^{1/\beta}} \left(\frac{z}{\eta}\right)^{1/\beta - 1} dz. \tag{2.47}$$

Expanding $e^{-\left(\frac{z}{\eta}\right)^{1/\beta}}$ in the series

$$e^{-\left(\frac{z}{\eta}\right)^{1/\beta}} = 1 - \frac{Z^{1/\beta}}{1!\eta^{1/\beta}} + \frac{Z^{2/\beta}}{2!\eta^{2/\beta}} - \frac{Z^{3/\beta}}{3!\eta^{3/\beta}} + \ldots + (-1)^i \frac{Z^{i/\beta}}{i!\eta^{i/\beta}} + \ldots, \tag{2.48}$$

from (2.47), by term-by-term integration, we obtain an expression for the reliability function at the first load in the form of a series

$$R_1 = 1 + \sum_{i=1}^{\infty} (-1)^i \frac{\Gamma(1 + i/\beta)}{i!\eta^{i/\beta}} \tag{2.49}$$

Using the safety factor for the average $\overline{K} = \frac{\overline{P}_\Pi}{\overline{P}_H}$, one can transform the series (2.49) to the form

$$R_1 = 1 - \frac{\Gamma(1 + b/b_H)}{1!(\theta\overline{K})^{b_\Pi}} + \frac{\Gamma(1 + 2b_\Pi/b_H)}{2!(\theta\overline{K})^{2b_\Pi}} - \frac{\Gamma(1 + 3b_\Pi/b_H)}{3!(\theta\overline{K})^{3b_\Pi}} + \ldots \tag{2.50}$$

where $\theta = \frac{\Gamma(1+1/b_H)}{\Gamma(1+1/b_\Pi)}$.

We also consider the well-known representation [19] in the form of a convergent power series of the relation

$$\frac{(\theta\overline{K})^{b_\Pi}}{(\theta\overline{K})^{b_\Pi} + \Gamma(1 + b/b_H)} = 1 - \frac{\Gamma(1 + b/b_H)}{(\theta\overline{K})^{b_\Pi}} + \frac{\Gamma^2(1 + 2b_\Pi/b_H)}{(\theta\overline{K})^{2b_\Pi}} - \frac{\Gamma^3(1 + 3b_\Pi/b_H)}{(\theta\overline{K})^{3b_\Pi}} + \ldots \tag{2.51}$$

Given that in real engineering calculations $(\theta\overline{K})^{b_\Pi}$ is large numbers, the reliability function R_1 should be close to unity. Under these assumptions, an approximate relationship can be proposed to estimate the reliability function at the first load on the basis of comparison of the first terms of the series (2.50) and (2.51):

$$R_1 = \frac{(\theta\overline{K})^{b_\Pi}}{(\theta\overline{K})^{b_\Pi} + \Gamma(1 + b/b_H)} \tag{2.52}$$

In the particular case for $b_H = b_\Pi = b$, from (2.52) we obtain the exact formula (2.26).

The probabilistically justified values of the safety factor can be determined from the expression that follows from (2.52)

$$\overline{K}(\gamma) = \frac{\Gamma(1 + 1/b_\Pi)}{\Gamma(1 + 1/b_H)} \left[\frac{\gamma\,\Gamma(1 + b/b_H)}{1 - \gamma}\right]^{1/b_\Pi}. \tag{2.53}$$

Comparison of the results obtained by formulas (2.52) and (2.53) with calculations using numerical integration confirms their rather high accuracy for values

Table 2.3 Probabilistically justified safety factor for the first (unit) load

$b_{\text{н}}$	$V_{\text{н}}$	Reliability function γ			
		0.99	0.999	0.9999	0.99999
12.15	0.1	$\frac{1.460}{1.183}$	1.766	2.134	2.579
5.797	0.2	$\frac{1.612}{1.406}$	1.950	2.357	2.849
3.713	0.3	$\frac{1.850}{1.672}$	2.237	2.704	3.268

$R_1 = \gamma \geq 0.99$ and $b_{\Pi}/b_{\text{H}} > 0.3$. With the use of (2.53) at $b_{\text{п}} = 12.15$ ($V_{\text{п}} = 0.1$), the values of the safety factor are given in Table. 2.3.

The data given for comparison in Table 2.3 at $\gamma = 0.99$ under the bar correspond to the case $V_{\Pi} = 0$. They are taken from Table 1.3 at $m = 1$ and show how much you can reduce the safety factor without changing the reliability level, if you eliminate the random scattering of the limit load of the elements.

2.4 Models Based on Frechet Distribution

Let's consider the case where the random load and the time-constant limit load are distributed according to the Frechet law (1.8). Then the load distribution functions and limit load functions are as follows:

$$F(P) = \exp\left[-\left(\frac{h_{\text{H}}}{P}\right)^{\rho_{\text{H}}}\right]; \quad G(P) = \exp\left[-\left(\frac{h_{\Pi}}{P}\right)^{\rho_{\Pi}}\right]. \tag{2.54}$$

When changing to unit distributions, we use the inverse distribution functions of the limit load

$$\psi(G) = \frac{h_{\Pi}}{\left(\ln \frac{1}{G}\right)^{1/\rho_{\Pi}}} \tag{2.55}$$

and extreme loads

$$\varphi(F) = \frac{h_{\text{H}}}{\left(\ln \frac{1}{F}\right)^{1/\rho_{\text{H}}}}. \tag{2.56}$$

Using (2.15) and (2.17), we obtain the corresponding functions of the unit load distribution

$$F_1(G) = \exp\left[-\left(\frac{h_{\text{H}}}{h_{\Pi}}\right)^{\rho_{\text{H}}}\left(\ln \frac{1}{G}\right)^{\rho_{\text{H}}/\rho_{\Pi}}\right] \tag{2.57}$$

and a unit distribution of the limit load

$$G_1(F) = \exp\left[-\left(\frac{h_\Pi}{h_H}\right)^{\rho_\Pi}\left(\ln\frac{1}{F}\right)^{\rho_\Pi/\rho_H}\right]. \tag{2.58}$$

In the special case, when the variation coefficients are equal and the random load and limit load are similar, then the form parameters are the same $\rho_H = \rho_\Pi = \rho$, and the safety factor according to the mean values is $\overline{K} = \frac{h_\Pi}{h_H}$. Then it follows from (2.57) that the function of the unit load distribution takes the form

$$F_1(G) = G^{1/\overline{K}^\rho}. \tag{2.59}$$

Using (2.19) and (2.59), we obtain an analytic expression for the reliability function of an element at m-times extreme load:

$$R_m = \int_0^1 G^{m/\overline{K}^\rho}\, dG = \frac{\overline{K}^\rho}{\overline{K}^\rho + m}. \tag{2.60}$$

Hence, it follows the formula for determining the probabilistically valid safety factors for the mean, which provide a given value γ of the reliability function with m-fold loading:

$$\overline{K}(\gamma, m) = \left(\frac{\gamma}{1-\gamma}\right)^{1/\rho} m^{1/\rho}. \tag{2.61}$$

As in the case of the Weibull distribution (see (2.34)), it follows from (2.60) that the number of loads prior to a sudden failure does not obey the geometric distribution in view of the fact that

$$R_m = \frac{\overline{K}^\rho}{\overline{K}^\rho + m} > \left(\frac{\overline{K}^\rho}{\overline{K}^\rho + 1}\right)^m = R_1^m, \text{ when } m = 2, 3, \ldots$$

In the general case, if the variation coefficients of the load and limit load are different and $\rho_H \neq \rho_\Pi$ it follows from (2.57) and (2.19) that the expression by which the reliability function can be determined by numerical integration, takes the form

$$R_m = \int_0^1 \exp\left[-m\mu\left(\ln\frac{1}{G}\right)^\alpha\right] dG, \tag{2.62}$$

where $\mu = \left(\frac{h_H}{h_\Pi}\right)^{\rho_H}$; $\alpha = \rho_H/\rho_\Pi$.

It is possible from here to obtain a similar equation to determine the values of the parameter $\mu_\gamma(\alpha)$, which at $m = 1$ correspond to the set values γ for the reliability function

$$\int_0^1 \exp\left[-\mu_\gamma(\alpha)\left(\ln\frac{1}{G}\right)^\alpha\right] dG = \gamma. \tag{2.63}$$

Table 2.4 illustrates some equations of a number of values α obtained by the numerical solution (2.63) and multiplied by 10^3 values $\mu_\gamma(\alpha)$, corresponding to the reliability function: $\gamma = 0.99$; 0.999 and 0.9999.

Proceeding from (2.62) and (2.63), expressions can be obtained to determine the probabilistically justified value of the safety factor according to the mean ones

$$\begin{aligned}\overline{K}(\gamma, m) &= \left(\frac{m}{\mu_\gamma(\alpha)}\right)^{1/\rho_H} \frac{\Gamma(1 - 1/\rho_\Pi)}{\Gamma(1 - 1/\rho_H)} \\ &= \left(\frac{m}{\mu_\gamma(\alpha)}\right)^{1/\rho_H} \frac{\alpha \cdot \sin\left(\frac{\pi}{\rho_H}\right)\Gamma(1 + 1/\rho_H)}{\sin\left(\frac{\pi}{\rho_\Pi}\right)\Gamma(1 + 1/\rho_\Pi)}.\end{aligned} \tag{2.64}$$

In this case, when there is no or not taken into account the random dispersion of the limit load, i.e. $V_\Pi \to 0$, $p_\Pi \to \infty$, $\alpha \to 0$, the expression (2.64) takes the form

$$\overline{K}(\gamma, m) = \left(\frac{m}{\ln 1/\gamma}\right)^{1/\rho_H} \frac{\rho_H \cdot \sin\left(\frac{\pi}{\rho_H}\right)\Gamma(1 + 1/\rho_H)}{\pi}. \tag{2.65}$$

Using the data on the parameter $\mu_\gamma(\alpha)$ given in Table 2.4, it is possible to calculate the probabilistically justified value of the safety factor for repeated extreme load. So, if when being designed, it is given that $\gamma = 0.99$, $m = 10$; $V_H = 0.2(\rho_H = 7.26)$, $\alpha = 0.5(\rho_\Pi = 14.52$, $V_\Pi \approx 0.093)$, then selecting from Table 2.4 $\mu_{0.99}(0.5) = 0.01135704$, with the help of the formula (2.64) we obtain:

$$\overline{K}(0.99; 10) = \left(\frac{10}{0.01135704}\right)^{1/7.26} \frac{0.5 \cdot \sin\left(\frac{\pi}{7.26}\right)\Gamma(1 + 1/7.26)}{\sin\left(\frac{\pi}{14.52}\right)\Gamma(1 + 1/14.52)} 1 = 2.415.$$

Table 2.5 gives some values of probabilistically justified safety factors according to the mean values, their calculation is based on the use of the data in Table. 2.4, i.e. on the Frechet distribution model. The comparison of Table 2.5 data with the corresponding data of Table 1.3 gives an idea of the degree of influence on the random dispersion of the limit load, specified during the design.

Table 2.4 Parameter indicators $\mu_\gamma(\alpha) \times 10^3$

$\alpha = {}^{\rho_H}/_{\rho_\Pi}$	Reliability function γ		
	0.99	0.999	0.9999
0	10.05034	1.000500	0.100005
0.1	10.56478	1.051644	0.105116
0.2	10.94895	1.089698	0.108918
0.3	11.20513	1.114908	0.111435
0.4	11.33882	1.127833	0.112723
0.5	11.35704	1.129276	0.112863
0.6	11.27022	1.120215	0.111952
0.7	11.08985	1.101753	0.110101
0.8	10.82657	1.075066	0.107427
0.9	10.49344	1.041028	0.104018
1.0	10.10101	1.001001	0.100010
1.1	9.66328	0.95717	0.095622
1.2	9.18947	0.909373	0.090838

Table 2.5 Probabilistically justified safety factors according to the mean values

	Variation coefficients of loads and limit load			
	$V_H = V_\Pi = 0.1$		$V_H = 0.2;\ V_\Pi = 0.093$	
Number of loadings m	$\gamma = 0.99$	$\gamma = 0.999$	$\gamma = 0.99$	$\gamma = 0.999$
1	1.401	1.661	1.758	2.417
10	1.659	1.966	2.415	3.318
50	1.868	2.213	2.938	4.142

2.5 Discrete Distribution of the Number of Loads to Failure

Earlier it was shown that for a deterministic constant limiting level of the limit load of an element, an integer number of random loads with a sudden failure has a discrete one-parameter geometric distribution with a probability function of (1.27). However, if the limit load of each element is a random fixed in time variable, then the distribution of the loading number to failure is not geometric under multiple random loading. This is evidenced by the expressions for the reliability function (2.34) and (2.60), obtained under the assumption that the load and limit load are random and have the Weibull distribution or are subject to the Frechet law. It is also assumed that the limit load and load are independent of each other, but similar random quantities, i. e. they have the same distribution law and the same variation coefficients.

It makes sense to consider the distinguishing features of the geometric distribution from the discrete distributions of the number of loads to failure, which correspond to the cases of distributions of similar load quantities and limit load according to the Weibull and Frechet laws. If the Weibull distribution takes place, then it follows from (2.26) that the probability of failure at the first load

$$Q_1 = 1 - R_1 = \frac{1}{\overline{K}^b + 1},$$

where $\overline{K}^b = \frac{1-Q_1}{Q_1}$.

Substitution of this quantity into (2.34) gives an expression for the reliability function for m-fold loading in the form of

$$R_m = \frac{(1-Q_1)^m m!}{\prod_{i=1}^{m} [(1-Q_1)i + Q_1]}; \quad m = 1, 2, \ldots \tag{2.66}$$

which should be compared with the corresponding expression (1.26) for the geometric distribution.

Using (2.66), we can obtain the probability function of the discrete distribution under consideration, depending only on one parameter Q_1:

$$Q_m = R_{m-1} - R_m = \frac{Q_1(1-Q_1)^{m-1}(m-1)!}{\prod_{i=1}^{m} [(1-Q_1)i + Q_1]}; \quad m = 1, 2, \ldots \tag{2.67}$$

This function determines the value of the failure probability specifically at the mth extreme load and is a discrete analog of the density distribution of the number of loads to failure.

It is easy to see that the normalization condition is satisfied at $0 < Q_1 < 1$ for (2.67) and the series $\sum_{m=1}^{\infty} Q_m$ converges to unity. A verification of the convergence of this series can be carried out with the Raabe criterion [19], in accordance with which the convergent series satisfies the condition:

$$\lim_{m \to \infty} \left[m \left(\frac{Q_m}{Q_{m+1}} - 1 \right) \right] = \lim_{m \to \infty} \left[m \left(\frac{(1-Q_1)m + 1}{(1-Q_1)m} - 1 \right) \right] = \frac{1}{1-Q_1} > 1,$$

if $0 < Q_1 < 1$.

The mean value of the distribution (2.67) is determined by the sum of the series

$$M = \sum_{m=1}^{\infty} \frac{Q_1(1-Q_1)^{m-1} m!}{\prod_{i=1}^{m} [(1-Q_1)i + Q_1]}. \tag{2.68}$$

The condition for the convergence of the series (2.68) can also be obtained with the Raabe test, by determining the value of the limit.

$$\lim_{m \to \infty} \left[m \left(\frac{(1 - Q_1)\, m + 1}{(1 - Q_1)\,(m + 1)} - 1 \right) \right] = \frac{Q_1}{1 - Q_1} > 1, \quad \text{when } Q_1 > 0.5.$$

Consequently, unlike the geometric distribution, the mean distribution (2.67) has a finite value only under the condition that the probability of failure is within the interval $0.5 < Q_1 < 1$ at the first load Q_1. In practice, this probability of failure in machinery elements is unacceptable. Then

$$M = \frac{Q_1}{2Q_1 - 1}. \tag{2.69}$$

The discrete distribution (2.67) was not found in the well-known reference books [12, 15], and in [20] this distribution is called hypo-geometric.

The probability function of the hypo-geometric distribution (2.67) differs from the corresponding characteristic (1.27) of the geometric distribution by the presence of a factor $\frac{(m-1)!}{\prod_{i=1}^{m} [(1-Q_1)i + Q_1]}$ that is less than unity for small values $m > 1$, but with an increase of m it begins to exceed unity and increases the probability values Q_m in comparison with the geometric distribution.

An important characteristic of the hypo-geometric distribution is the conditional distribution function (risk function), which is defined as the ratio

$$\lambda_m = \frac{Q_m}{R_{m-1}} = \frac{Q_1}{(1 - Q_1)m + Q_1}; \quad m = 1, 2, \ldots \tag{2.70}$$

This function allows for each next mth loading to determine the conditional probability of failure of an object that, having withstood $m - 1$ loading, did not fail. As follows from (2.70), the risk function of the hypo-geometric distribution is a monotonically decreasing function of the m number of loads, and this in principle differs from the geometric distribution, which λ_m is equal to a constant value Q_1. Reducing the risk of failure due to preloading in a number of cases justifies the advisability of using such a method of ensuring the reliability of elements that have a random dispersion of the limit load.

Another version of the discrete distribution of the number of loads to failure can be obtained on the basis of the expression (2.60) for the reliability function in the case of a similar load and limit load, distributed according to the Frechet law. Expressing $\overline{K}^{\rho}$ with Q_1 and substituting into (2.60), we obtain an expression for the reliability function under m-fold loading in the following form:

$$R_m = \frac{1 - Q_1}{1 + (m - 1)Q_1}; \quad m = 1, 2, \ldots \tag{2.71}$$

On the basis of (2.71), we can obtain the remaining characteristics of this distribution, which are given in Table 2.6.

Table 2.6 The main characteristics of discrete distributions number of loads to failure

Kind of distribution	Geometric	Quasi-geometric	Hypo-geometric
Reliability function	$(1-Q_1)^m$	$\frac{1-Q_1}{1+(m-1)Q_1}$	$\frac{(1-Q_1)^m m!}{\prod_{i=1}^{m}[(1-Q_1)i+Q_1]}$
Function of probability of failure	$Q_1(1-Q_1)^{m-1}$	$\frac{Q_1(1-Q_1)}{[1+(m-1)Q_1][1+(m-2)Q_1]}$	$\frac{Q_1(1-Q_1)^{m-1}(m-1)!}{\prod_{i=1}^{m}[(1-Q_1)i+Q_1]}$
Function of failure risk	Q_1	$\frac{Q_1}{(m-1)Q_1+1}$	$\frac{Q_1}{(1-Q_1)m+Q_1}$
Mean distribution	$\frac{1}{Q_1}$	–	$\frac{Q_1}{2Q_1-1}$; when $0.5 < Q_1 < 1$

In view of the greater than the hypo-geometric proximity of this distribution to the geometric and its absence in [12, 15], it is called quasigeometric in the table. The convergence of the sum of the series $\sum_{m=1}^{\infty} Q_m$ to unity is verified with the help of the Raabe attribute by the value of the limit

$$\lim_{m\to\infty}\left[m\left(\frac{Q_m}{Q_{m+1}}-1\right)\right]=\lim_{m\to\infty}\left[m\left(\frac{1+mQ_1}{1+mQ_1-2Q_1}-1\right)\right]=2>1,$$

i.e. the convergence condition is satisfied.

The mean of the quasi-geometric distribution is a sum of the series

$$M=\sum_{m=1}^{\infty}\frac{Q_1(1-Q_1)m}{[1+(m-1)Q_1][1+(m-2)Q_1]}. \tag{2.72}$$

An analysis of the convergence of the series (2.72) is carried out with the Gauss test [19]. In accordance with this criterion, we consider the ratio of the terms of the series (2.72), which is represented in the form

$$\frac{a_{k+1}}{a_k}=\frac{m^2+m\left(\frac{1}{Q_1}-1\right)+\frac{1}{Q_1}-2}{m^2+\frac{1}{Q_1}m}.$$

Since the difference of the coefficients for m is: $\frac{1}{Q_1}-\left(\frac{1}{Q_1}-1\right)=1$, then the series (2.72) diverges and there is no finite mean for the quasi-geometric distribution. Theoretically, this means that in cases of quasi-geometric and hypo-geometric (at $Q_1 \le 0.5$) distributions, some of the objects cannot be brought to failure for any finite number of loads.

The risk function $\lambda_m=\frac{Q_1}{(m-1)Q_1+1}$ of a quasi-geometric distribution decreases monotonically with increasing number of loads, tending to zero.

Figure 2.1 and 2.2 represent the graphs of probability functions and risk functions of discrete distributions of the number of loads to failure with a parameter value $Q_1 = 0.1$.

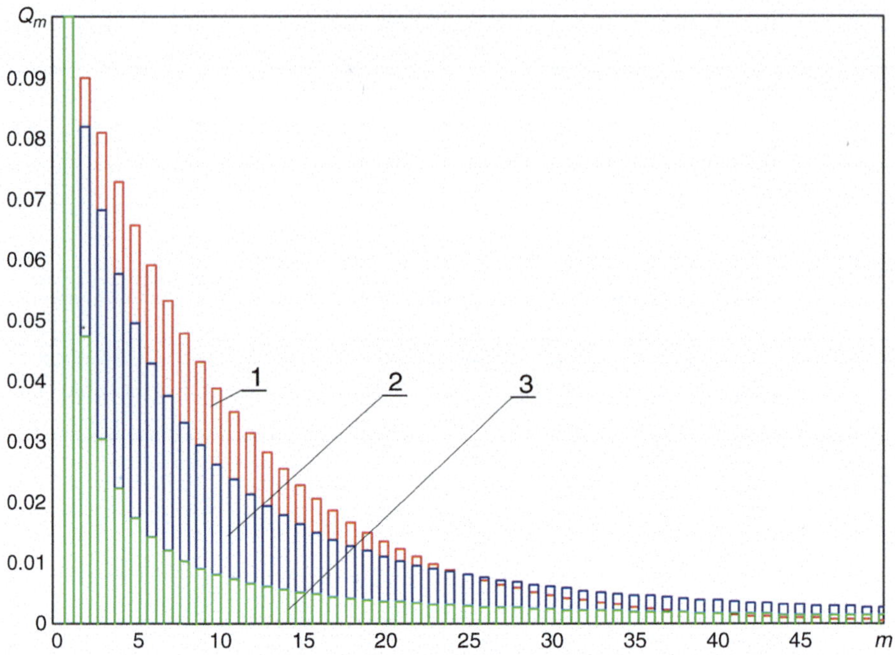

Fig. 2.1 Graphs of probability functions of discrete distributions of the number of loads to failure: *1*, geometric; *2*, quasi-geometric; *3*, hypo-geometric

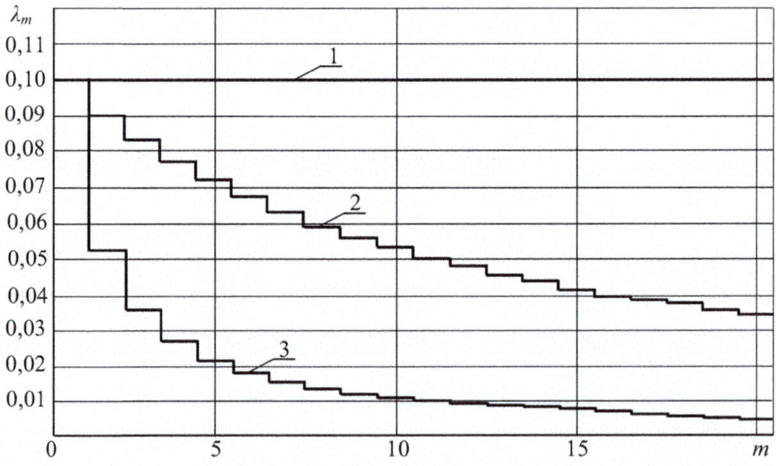

Fig. 2.2 Graphs of risk functions of discrete distributions: *1*—geometric; *2*—quasi-geometric; *3*—hypo-geometric

It should be noted that it is the random scattering of the limit load of the elements that results in the distribution of the number of loads before sudden failure acquires properties different from the properties of the geometric distribution.

The qualitative and quantitative differences between the hypo-geometric and quasi-geometric distributions from the geometric distributions are very important and this is graphically illustrated by the graphs shown in Fig. 2.1 and 2.2. The graphs of the failure probability functions in Fig. 2.1 show that for small values of the number of loads $m > 1$, the geometric probability distribution is higher, but then the probability value Q_m decreases much faster and becomes smaller than for quasi-geometric and hypo-geometric distributions. The most significant differences are manifested in the behavior of the risk function (Fig. 2.2), which for the geometric distribution remains constant and equal to the probability of failure at the first load Q_1, and in quasi-geometric and especially in hypo-geometric distributions, the risk of failure decreases with the increase in the number of loads to values much lower, than Q_1. Another distinguishing feature of discrete distributions associated with scattering of the limit load is the absence of a finite average number of loads to failure.

Therefore, contrary to the generally accepted approach [9, 22, 48] with the forecast of mechanical reliability by sudden failures in conditions of random dispersion of the limit load of elements, the use of the geometric distribution cannot be considered to be justified.

2.6 Models Based on Log-Logistic Distribution

Let's consider the case where the load and limit load of the elements have a log-logistic distribution with the following functions:

$$F(P) = \frac{\left(P/C_{\text{H}}\right)^{V_{\text{H}}}}{1 + \left(P/C_{\text{H}}\right)^{V_{\text{H}}}}; \quad G(P) = \frac{\left(P/C_{\Pi}\right)^{V_{\Pi}}}{1 + \left(P/C_{\Pi}\right)^{V_{\Pi}}}. \tag{2.73}$$

The inverse distribution function of the limit load in this case takes the form

$$\psi(G) = \frac{C_{\Pi} G^{1/v_{\Pi}}}{(1 - G)^{1/v_{\Pi}}}. \tag{2.74}$$

Then the function of the unit load distribution is determined from the expression:

$$F_1(G) = \frac{G^{\delta}}{q(1 - G)^{\delta} + G^{\delta}}, \tag{2.75}$$

where $q = \left(\frac{C_{\text{H}}}{C_{\Pi}}\right)^{V_{\text{H}}}$; $\nu_{\text{H}}/\nu_{\Pi} = \delta$

The reliability function of the element at the first load in accordance with (2.16) is calculated from the formula:

$$R_1 = \int_0^1 \frac{G^\delta}{q(1-G)^\delta + G^\delta} dG. \quad (2.76)$$

In the case of similar load and carrying capacity, when $\nu_H = \nu_\Pi = \nu$ and $\delta = 1$, and the safety factor for the average $\overline{K} = \frac{C_\Pi}{C_H}$, based on (2.76), we obtain

$$R_1 = \int_0^1 \frac{\overline{K}^\nu G}{1 + G(\overline{K}^\nu - 1)} dG. \quad (2.77)$$

If $\overline{K} = 1$, then it follows from (2.77) that

$$R_1 = \int_0^1 G dG = \frac{1}{2}.$$

In the case when $\overline{K} \neq 1$, the change of variable $x = \frac{\overline{K}^\nu G}{1+G(\overline{K}^\nu - G)}$ leads the integral (2.77) to the tabular [21] form:

$$R_1 = \frac{\overline{K}^\nu}{(\overline{K}^\nu - 1)^2} \int_0^1 \frac{xdx}{\left(x + \frac{\overline{K}^\nu}{1-\overline{K}^\nu}\right)^2}. \quad (2.78)$$

After integration, we obtain an analytical expression for the reliability function at the first load:

$$R_1 = \begin{cases} \frac{\overline{K}^\nu}{(\overline{K}^\nu - 1)^2}\left[\overline{K}^\nu - 1 - \ln \overline{K}^\nu\right]; \text{ when } \overline{K} \neq 1; \\ \frac{1}{2}; \text{ when } \overline{K} = 1. \end{cases} \quad (2.79)$$

Proceeding from the expression (2.79) valid for such random values of load and limit load of the elements, we can determine the value of the safety factor corresponding to a given reliability function at the first load. Such a safety factor is a solution of the equation

$$\frac{\overline{K}^\nu(\gamma)}{(\overline{K}^\nu(\gamma) - 1)^2}\left[\overline{K}^\nu(\gamma) - 1 - \ln \overline{K}^\nu(\gamma)\right] = \gamma. \quad (2.80)$$

In the absence of similarity, when $\nu_{\mathrm{H}} \neq \nu_{\Pi}$, we should use the general expression (2.76), defining the value of the parameter q_γ, which is the root of equation

$$\int_0^1 \frac{G^\delta}{q_\gamma (1+G)^\delta + G^\delta} dG = \gamma. \tag{2.81}$$

Then we calculate the corresponding safety factor:

$$\overline{K}(\gamma) = \frac{\delta \cdot \sin \frac{\pi}{\nu_{\mathrm{H}}}}{q_\gamma^{1/\nu_{\mathrm{H}}} \cdot \sin \frac{\delta_\pi}{\nu_{\mathrm{H}}}}. \tag{2.82}$$

In this way, with the help of (2.80), (2.81), and (2.82), the safety factors were determined provided that the variation coefficient of the limit load is everywhere assumed to be $\nu_\Pi = 0,1$ and the coefficient of load variation ν_{H} takes a number of specified values. Table 2.7 contains these data and gives for comparison safety factors calculated with the formulas (2.3) and (2.7), corresponding to the normal and log-normal distributions.

The data presented in Table 2.7, and also in Tables 2.3 and 2.5 indicate a significant influence of the type of load distributions and limit load on the value of the probabilistically justified safety factor. In particular, it can be noted that the use of the normal distribution in most cases leads to lower values of the probabilistically justified safety factor in comparison with the other distributions considered. This is not taken into account in the recommendations that we can sometimes encounter in the literature on machine reliability [9, 23, 39, 49].

The reliability function at m-fold loading in accordance with (2.19) and (2.75) can be determined from the expression:

Table 2.7 Probabilistically justified safety factor at the first (unit) load

		Reliability function γ			
Kind ofdistribution	ν_{H}	0.99	0.999	0.9999	0.99999
Normal	0.1	1.40	1.59	1.77	1.95
	0.2	1.60	1.84	2.07	2.30
	0.3	1.82	2.14	2.44	2.73
Lognormal	0.1	1.39	1.55	1.69	1.83
	0.2	1.65	1.96	2.25	2.54
	0.3	1.98	2.52	3.05	3.62
Logarithmic logistic	0.1	1.41	1.64	1.89	2.17
	0.2	1.70	2.18	2.79	3.58
	0.3	2.05	2.94	4.24	6.08

$$R_m = \int_0^1 \left(\frac{G^\delta}{q(1-G)^\delta + G^\delta} \right)^m dG. \tag{2.83}$$

If the load and limit load are similar random variables ($\delta = 1$), then we get from (2.83) that

$$R_m = \int_0^1 \left(\frac{\overline{K}^\nu G}{1 + G(\overline{K}^\nu - G)} \right)^m dG. \tag{2.84}$$

For $\overline{K} = 1$ we have the formula for a particular case:

$$R_m = \int_0^1 G^m dG = \frac{1}{m+1}. \tag{2.85}$$

After transition to (2.84) to the variable x (see (2.78)), we obtain

$$R_m = \begin{cases} \dfrac{\overline{K}^\nu}{(\overline{K}^\nu - 1)^2} \displaystyle\int_0^1 \dfrac{x^m dx}{\left(x + \frac{\overline{K}^\nu}{1-\overline{K}^\nu}\right)^2}; & \text{when } \overline{K} \neq 1; \\ \dfrac{1}{m+1}; & \text{when } \overline{K} = 1. \end{cases} \tag{2.86}$$

The integral in (2.86) refers to the tabular [21] and, as a result of its calculation, the expression for the reliability function takes on the following analytical form:

$$R_m = \begin{cases} \dfrac{\overline{K}^\nu}{(\overline{K}^\nu - 1)^2} \Big[(-1)^m \overline{K}^{\nu(m-1)} \Big((1-\overline{K}^\nu)^{2-m} + m(1-\overline{K}^\nu)^{1-m} \ln \overline{K}^\nu \Big) \\ + \displaystyle\sum_{i=1}^{m-1} (-1)^{i+1} \dfrac{i}{(m-i)} \left(\dfrac{\overline{K}^\nu}{1-\overline{K}^\nu} \right)^{i-1} \Big]; & \text{when } \overline{K} \neq 1; \\ \dfrac{1}{m+1}; \quad \text{when } \overline{K} = 1. \end{cases} \tag{2.87}$$

Table 2.8 gives formulas corresponding to small values of the number of loads. For large values $m \geq 5$, these formulas become cumbersome and computer calculations are more conveniently carried out using the general expressions (2.84) or (2.86) with numerical integration.

Table 2.8 Formulas to calculate reliability function

Number of loadings m	Reliability function
1	$R_1 = \begin{cases} \dfrac{\overline{K}^{\nu}}{(\overline{K}^{\nu} - 1)^2}\left[\overline{K}^{\nu} - 1 - \ln \overline{K}^{\nu}\right]; \text{ when } \overline{K} \neq 1; \\ \dfrac{1}{2}; \text{ when } \overline{K} = 1. \end{cases}$
2	$R_2 = \begin{cases} \dfrac{\overline{K}^{\nu}}{(\overline{K}^{\nu} - 1)^2}\left(\overline{K}^{\nu} + 1 - \dfrac{2\overline{K}^{\nu}}{\overline{K}^{\nu} - 1} \ln \overline{K}^{\nu}\right); \text{ when } \overline{K} \neq 1; \\ \dfrac{1}{3}; \text{ when } \overline{K} = 1. \end{cases}$
3	$R_3 = \begin{cases} \dfrac{\overline{K}^{\nu}}{(\overline{K}^{\nu} - 1)^2}\left(\dfrac{1}{2} + \dfrac{\overline{K}^{\nu}(\overline{K}^{\nu} + 2)}{\overline{K}^{\nu} - 1} - \dfrac{3\overline{K}^{2\nu}}{(\overline{K}^{\nu} - 1)^2} \ln \overline{K}^{\nu}\right); \text{ when } \overline{K} \neq 1; \\ \dfrac{1}{4}; \text{ when } \overline{K} = 1. \end{cases}$
4	$R_4 = \begin{cases} \dfrac{\overline{K}^{\nu}}{(\overline{K}^{\nu} - 1)^2}\left(\dfrac{1}{3} + \dfrac{\overline{K}^{\nu}}{\overline{K}^{\nu} - 1} + \dfrac{\overline{K}^{2\nu}(\overline{K}^{\nu} + 3)}{(1 - \overline{K}^{\nu})^2} - \dfrac{4\overline{K}^{3\nu}}{(\overline{K}^{\nu} - 1)^3} \ln \overline{K}^{\nu}\right); \text{ when } \overline{K} \neq 1; \\ \dfrac{1}{5}; \text{ when } \overline{K} = 1. \end{cases}$

A numerical verification of the inequality $R_m > R_1^m$ in the case of a logarithmic logistic distribution shows that, as in the cases of the Weibull and Frechet distributions, this inequality holds. Consequently, in this case also the distribution of the number of loads before a sudden failure is not geometric. If we combine all the previous variants of distributions in a special case, when $\overline{K} = 1$ and, respectively, the probability of failure at the first load $Q_1 = 0.5$, then the same expression for the reliability function follows from (2.34), (2.60) and (2.85):

$$R_m = \frac{1}{m+1}; \quad m = 1, 2, \ldots \tag{2.88}$$

Failure probability function under m loading

$$Q_m = R_{m-1} - R_m = \frac{1}{m(m+1)}; \quad m = 1, 2, \ldots \tag{2.89}$$

The normalization condition for the distribution (2.89) is satisfied, since the series $\sum_{m=1}^{\infty} \frac{1}{m(m+1)}$ converges [21] to unity. The finite mean of the distribution (2.89) does not exist, in view of the fact that the series $\sum_{m=1}^{\infty} \frac{1}{m+1}$ diverges. Failure Risk Function

$$\lambda_m = \frac{Q_m}{R_{m-1}} = \frac{1}{m+1}; \quad m = 1, 2, \ldots \tag{2.90}$$

decreases monotonically with increasing number of loads, which distinguishes the distribution (2.89) from the geometric distribution.

Let us further consider the estimation of probabilistically justified safety factors in the general case of m-fold loading, when the reliability function is determined by the expression (2.83).

If for a given value of the parameter δ and the known number of loads m by means of a numerical solution of equation

$$\int_0^1 \frac{G^{\delta m} dG}{\left[(1-G)^{\delta} \cdot q_{\gamma}(m) + G^{\delta}\right]^m} = \gamma \tag{2.91}$$

we want to define the value of the parameter $q_\gamma(m)$ corresponding to the specified value of the reliability function γ, then, in accordance with (2.82), the value of the probabilistically justified safety factor will be determined by the expression

$$\overline{K}(\gamma, m) = \frac{\delta \cdot \sin \frac{\pi}{\nu_H}}{\left(q_{\gamma}(m)\right)^{1/\nu_H} \cdot \sin \frac{\delta \cdot \pi}{\nu_H}}.$$

Table 2.9 Data on the dependence of the parameter $q_\gamma(m)$ on the number of loads

		m				
δ	γ	1	10	100	500	1000
1	0.999	0.125194	0.012092	0.00119831	0.000239461	0.0001197191
	0.9999	0.094550	0.013536	0.00135361	0.000270661	0.000135361
0.5088	0.999	0.6375	0.06363	0.0063649	0.001273	0.00063646
	0.9999	0.6355	0.06356	0.006355	0.0012711	0.00063549
0.3488	0.999	0.8260	0.08147	0.008147	0.0016293	0.0008147
	0.9999	0.8139	0.08137	0.008137	0.0016283	0.0008137
0.3042	0.999	0.85735	0.085675	0.0085675	0.0017135	0.00085675
	0.9999	0.8561	0.085585	0.0085585	0.0017123	0.00085585

Table 2.9 illustrates the data for a number of values of the parameter δ on the variation $q_\gamma(m)$ depending on the number of loads m for reliability function values of 0.999 and 0.9999. To determine the value $q_\gamma(m)$, the number given in the table is divided by 103 if γ = 0.999 or divided by 104 if γ = 0.9999.

The values of δ in Table 2.9, which are not equal to unity, are chosen so that when the variation coefficient of the limit load $V_П = 0.1$ ($\nu_П = 18.25$) they correspond to the values of the load variation coefficient $V_Н = 0.2$, 0.3 and 0.35.

Using Tables 2.9 and 1.1, it is possible to calculate the value of the probabilistically justified safety factor.

So, for example, for $V_П = 0.1$ and $V_Н = 0.2$ the value is $\delta = {}^{9.28}\!/_{18.25} = 0.5088$. We find using Table 2.9 the value $q_{0.999}(100) = 0.0063649 \cdot 10^{-3}$ for γ = 0.999 and m = 100 and we calculate the safety factor.

$$\overline{K}(0.999; 100) = \frac{0.5088 \cdot \sin \frac{\pi}{9.28}}{\left(0.0063649 \cdot 10^{-3}\right)^{1/9.28} \sin \frac{0.5088 \cdot \pi}{9.28}} = 3.578.$$

When using Table 2.9, you can interpolate the dependence $q_\gamma(m)$ on the number of loads using an expression $q_\gamma(m) \approx {}^{10q_\gamma(10)}\!/_m$ that is true for $10 \le m \le 1000$. For instance, if you want to calculate the safety factor in the conditions of the previous variant, but with γ = 0.9999 and $m = 20$, then

$$q_{0.9999}(20) = {}^{10 \cdot 0.06356 \cdot 10^{-4}}\!/_{20} = 0.03178 \cdot 10^{-4}$$

and correspondingly

$$\overline{K}(0.9999; 20) = \frac{0.5088 \cdot \sin \frac{\pi}{9.28}}{\left(0.03178 \cdot 10^{-4}\right)^{1/9.28} \cdot \sin \frac{0.5088 \cdot \pi}{9.28}} = 3.856.$$

Comparison of the data shown in Table 2.7 and 2.9 leads to the conclusion that it is possible to calculate the probabilistically justified values of the safety factor $\overline{K}(\gamma, m)$ in the case of a log-logistic distribution according to the simplified formula:

$$\overline{K}(\gamma, m) = \overline{K}(\gamma) \cdot m^{1/\nu_{\text{H}}}, \tag{2.92}$$

where $\overline{K}(\gamma)$ is a safety factor corresponding to the specified reliability function at the first load (see Table 2.7).

So, if we accept the terms of the example considered $V_{\text{H}} = 0.2$; ($\nu_{\text{H}} = 9.28$), $\gamma = 0.999$ and $m = 100$, then from Table 2.7 we find $\overline{K}(0.999) = 2.18$ and using formula (2.92), we calculate

$$\overline{K}(0.999; 100) = 2.18 \cdot 100^{1/9.28} = 3.58.$$

This result practically coincides with the result obtained from the data of Table 2.9 in the previous example.

Chapter 3
Management of Elements Reliability

3.1 Models Based on Combinations of Distributions

Consider the models in which the load and limit load do not obey the same distribution laws. We will use some variants of combinations of different types of load distributions and limit load. If the load is distributed according to the Frechet law with a distribution function of the form (2.54), and the limit load has a log-logistic distribution (2.73), then the corresponding function of the unit load distribution takes the form:

$$F_1(G) = \exp\left[-\left(\frac{h_{\text{H}}}{C_{\Pi}}\right)^{\rho_{\text{H}}} \frac{(1-G)^{\rho/\nu_{\Pi}}}{G^{\rho_{\text{H}}/\nu_{\Pi}}}\right]. \tag{3.1}$$

Then the reliability function for an m-fold loading of an element is determined from the general expression:

$$R_m = \int_0^1 \exp\left[-m\left(\frac{h_H}{C_{\Pi}}\right)^{\rho_{\text{H}}} \frac{(1-G)^{\rho/\nu_{\Pi}}}{G^{\rho_{\text{H}}/\nu_{\Pi}}}\right] dG. \tag{3.2}$$

In the particular case, when $\rho_{\text{H}} = \nu_{\Pi}$ the load and limit load coefficients are interdependent, where $V_{\text{H}} < V_{\Pi}$. Then it follows from (3.2) that

$$R_m = \int_0^1 \exp\left[-m\left(\frac{\Gamma\left(1+1/\rho_{\text{H}}\right)}{\overline{K}}\right)^{\rho_{\text{H}}} \frac{(1-G)}{G}\right] dG. \tag{3.3}$$

O. Grynchenko, O. Alfyorov, *Mechanical Reliability*,
https://doi.org/10.1007/978-3-030-41564-8_3

After the change of variable $x = \frac{m\Gamma^{\rho_H}\left(1+1/\rho_H\right)}{\overline{K}^{\rho_H} \cdot G}$, the integral (3.3) takes the form

$$R_m = A_m \cdot e^{A_m} \int_{A_m}^{\infty} \frac{e^{-x}}{x^2} dx, \tag{3.4}$$

where $A_m = m\left(\frac{\Gamma\left(1+1/\rho_H\right)}{\overline{K}}\right)^{\rho_H}$.

As a result of integration (3.4), we obtain for the reliability function an analytic expression containing the integral exponential function:

$$R_m = 1 - A_m \cdot e^{A_m} \cdot E_1(A_m), \tag{3.5}$$

where $E_1(A_m) = \int_{A_m}^{\infty} \frac{e^{-x}}{x} dx$ is the value of the integral exponential function tabulated in [25, 46].

For $\overline{K}^{\rho_H} \geq m$ or $A_m \leq 1$ in the expression (3.5), we can use the polynomial approximation [46] to calculate the integral exponential function:

$$E_1(A_m) = -\ln A_m - C_o + a_1 A_m - a_2 A_m^2 + a_3 A_m^3 - a_4 A_m^4 + a_5 A_m^5, \tag{3.6}$$

where $C_o = 0.57722$; $a_1 = 0.99999$; $a_2 = 0.24991$; $a_3 = 0.05520$; $a_4 = 0.00976$; $a_5 = 0.00108$.

To estimate the probabilistically justified safety factors, it is expedient to use such values of the parameter $A_m(\gamma)$, which satisfy the equation

$$1 - A_m(\gamma) e^{A_m(\gamma)} E_1(A_m(\gamma)) = \gamma, \tag{3.7}$$

where γ is a specified value of the reliability function.

Then the probabilistically justified safety factors are determined from the expression

$$\overline{K}(\gamma, m) = \left(\frac{m}{A_m(\gamma)}\right)^{1/\rho_H} \Gamma\left(1 + 1/\rho_H\right). \tag{3.8}$$

Determination of the probabilistically justified safety factors in the case of a combination of the load distribution according to the Frechet law and the log-logistic distribution of the limit load will be considered in the example. Let the variation coefficient of the limit load be $V_\Pi = 0.1$. Then the parameter of the log-logistic distribution of the limit load is $V_\Pi = 18.25$ (see Table 1.1). Expression (3.5) is valid if the shape parameter of the load distribution according to the Frechet law is $\rho_H = 18.25$. This corresponds to the value of the variation coefficient of the load, calculated by the formula (1.10), which is $V_H \approx 0.0738$. The results of a numerical

Table 3.1 Probabilistically justified safety factors

Reliability function γ	Parameter $A_m(\gamma)$	Number of loadings m	Safety factor $\overline{K}(\gamma, m)$	Safety factors: Normal distribution	Safety factors: Lognormal
0.99	$1.725 \cdot 10^{-3}$	1	1.376	1.360	1.338
		10	1.561		
		50	1.705		
0.999	$0.118 \cdot 10^{-3}$	1	1.594	1.525	1.472
		10	1.809		
		50	1.975		
0.9999	$0.906 \cdot 10^{-5}$	1	1.835	1.684	1.590
		10	2.082		
		50	2.274		

solution of the eq. (3.7) for values $\gamma = 0.99$; 0.999 and 0.9999 are given in Table 3.1. The found values of the parameter $A_m(\gamma)$ were then used to determine the probabilistically justified safety factors $\overline{K}(\gamma, m)$ using the formula (3.8).

For comparison, it was also made (see Table 3.1) calculations of probabilistically justified safety factors $\overline{K}(\gamma)$ in the case of normal (p. 2.3) and log-normal (p. 2.7) load distributions and limit load with coincident probability values γ and variation coefficients V_{H} and V_{Π} with those that used in determining the values $\overline{K}(\gamma, m)$, i.e. at $V_{\text{H}} = 0.0738$ and $V_{\Pi} = 0.1$. As follows from Table 3.1, the higher values of the safety factors were obtained with the considered combination: The Frechet law is a log-logistic distribution. Consequently, this combination can be classified as unfavorable in the sense of reliability.

It is easy to see that in the particular case of the combination of distributions considered above, when the reliability function for m-fold loading is determined from expression (3.5), the following condition is satisfied: $R_m > R_1^m$. For example, if $m = 1$ and $r^{\rho_{\text{H}}}\left(1 + 1/\rho_{\text{H}}\right)/\overline{K}^{\rho_{\text{H}}} = 0,01$, then using tables [46] and formula (3.5), we obtain that $R_1 = 0.959215$. If $m = 50$, then $A_{50} = 0.5$ it also follows from (3.5) that $R_{50} = 0.53854 > R_1^{50} = 0.12468$. This confirms the illegitimacy of using the geometric distribution (2.26) in the analysis and prediction of reliability in the case of multiple loading, if the limit load of the elements has random scattering.

In addition, this indicates the possibility of managing reliability by conducting a series of preliminary random loads of the element prior to its operation. The number of such loads is m_{Π}, and the load size should have the distribution (2.54). Analogously to (2.35) in the case under consideration, the conditional reliability function of an element that has withstood the first m_{Π} preliminary loads without failure is determined from the expression:

$$R\left({}^{m}/_{m_{\Pi}}\right) = \frac{R_{m_{\Pi}+m}}{R_{m_{\Pi}}} = \frac{1 - A_{m_{\Pi}+m} e^{A_{m_{\Pi}+m}} E_1\left(A_{m_{\Pi}+m}\right)}{1 - A_{m_{\Pi}} e^{A_{m_{\Pi}}} E_1\left(A_{m_{\Pi}}\right)}. \tag{3.9}$$

The use of formulas (3.5) and (3.9) implies that for each series of loads the limit load of the loaded element remains fixed in time, i.e., it is determined only by the initial quality of the element. Consequently, the influence of any processes in the environment and in the material of the element on its mechanical characteristics is excluded.

Under the conditions of the example under consideration, at $m_{\Pi} = 0$, the reliability function of the element at the first load is $R(1/0) = R_1 = 0.959215$. But after one preliminary loading with the help of tables of the integral exponential function [25], using formula (3.9), we make a prediction that

$$R(1/1) = \frac{R_2}{R_1} = \frac{0.931550}{0.959215} = 0.97116.$$

By increasing the number of preloads to $m_{\Pi} = 22$, it is possible to increase the level of the predicted conditional reliability function of the elements at the first load after the "break-in", exceeding the value 0.99, which is sometimes adopted as a normative one:

$$R(1/22) = \frac{R_{23}}{R_{22}} = \frac{0.68186}{0.68600} = 0.99397.$$

However, it should be borne in mind that the number of elements that failed as a result of such break-in should be on average 31.4%.

Another variant of combinations of different load distributions and limit load, which allows us to obtain an analytical expression for the reliability function in a particular case, is a combination of the Weibull distribution for the load and the log-logistic distribution of the limit load.

Let the load have a Weibull distribution with a function $F(P)$ of the form (2.21), and the limit load $G(P)$ distribution function is represented in the form (2.73). In this case, the inverse distribution function of the limit load $\psi(P)$ has the form (2.74). After substitution $\psi(P)$ in $F(P)$, we get an expression for the function of the unit load distribution:

$$F_1(G) = 1 - \exp\left[-\frac{B \cdot G^{b_o}}{(1-G)^{b_o}}\right], \tag{3.10}$$

where the parameters of the unit distribution are related to the parameters of the initial load distributions (2.21) and the limit load (2.73) by the dependences

$$B = \left(\frac{C_{\Pi}}{a_{H}}\right)^{b_H}; \quad b_o = b_H/\nu_{\Pi}.$$

In accordance with (2.19), in the general case the reliability function of an element for m-fold load can be determined by numerical integration of expression

$$R_m = \int_0^1 \left(1 - \exp\left[-\frac{B \cdot G^{b_o}}{(1-G)^{b_o}}\right]\right)^m dG. \tag{3.11}$$

In this case, the safety factor is calculated by the formula

$$\overline{K} = \frac{C_{\Pi}\pi}{a_{\rm H}\nu_{\Pi}\Gamma\left(1 + {}^1\!/_{b_{\rm H}}\right) \cdot \sin {}^{\pi}\!/_{\nu_{\Pi}}}. \tag{3.12}$$

In the particular case when $b_{\rm H} = \nu_{\Pi}$ and $b_o = 1$, the unit distribution with the function (3.10) becomes a one-parameter one with a function of the form

$$F_1(G) = 1 - \exp\left[-\frac{B \cdot G}{1 - G}\right]. \tag{3.13}$$

At $m = 1$ the reliability function at the first load

$$R_1 = 1 - \int_0^1 \exp\left[-\frac{B \cdot G}{1 - G}\right] dG. \tag{3.14}$$

Making a change of variable $x = \frac{B}{1-G}$, from (3.14) we obtain

$$R_1 = 1 - B \cdot e^B \int_B^{\infty} \frac{e^{-x}}{x^2} dx. \tag{3.15}$$

Transforming (3.15) in analogy with (3.4) and (3.5), we obtain the expression containing the integral exponential function for the reliability function at the first load:

$$R_1 = B \cdot e^B E_1(B), \tag{3.16}$$

where $B = \left(\overline{K}{}^{\nu_{\Pi}}\!/_{\pi} \sin {}^{\pi}\!/_{\nu_{\Pi}} \Gamma\left(1 + {}^1\!/_{\nu_{\Pi}}\right)\right)^{V_{\Pi}}$.

Applying (3.16) for calculating R_1 at $B \geq 10$, one can use the approximation given in [46]:

$$R_1 = \frac{B^2 + c_1 B + c_2}{B^2 + e_1 B + e_2}, \tag{3.17}$$

where $c_1 = 4.0364;\ c_2 = 1.15198;\ e_1 = 5.03637;\ e_2 = 4.1916.$

Table 3.2 Probabilistically justified safety factors

Reliability function γ	$B(\gamma)$	Safety factors $\overline{K}(\gamma)$		
		Weibull distribution and log-logistic	Normal distribution	Lognormal
0.99	98.019	1.330	1.352	1.327
0.999	997.98	1.510	1.513	1.456
0.9999	9997.7	1.714	1.670	1.569

Using (3.17), we can calculate the parameter $B(\gamma)$ corresponding to a given probability $R_1 = \gamma$ by the formula:

$$B(\gamma) = \frac{e_1\gamma^{-C_1+}\sqrt{(e_1\gamma - c_1)^2 + 4(1-\gamma)(e_2\gamma - c_2)^2}}{2(1-\gamma)}, \tag{3.18}$$

and then determine a probabilistically justified safety factor

$$\overline{K}(\gamma) = \frac{\pi B^{1/v_{\Pi}}(\gamma)}{v_{\Pi} \sin \pi/_{v_{\Pi}} \Gamma\left(1 + 1/_{v_{\Pi}}\right)}. \tag{3.19}$$

Let us consider an example of determining probabilistically justified safety factor with the help of expressions (3.18) and (3.19). Suppose we have a case where a variation coefficient of the limit load $V_{\Pi} = 0.1$, which corresponds to the value of the parameter of the form of the log-logistic distribution $v_{\Pi} = 18.25$ (see Table 1.1). The use of formulas (3.18) and (3.19) assumes that the parameter of the Weibull distribution form of the load is $b_{\mathrm{H}} = v_{\Pi}$. Therefore, the corresponding variation coefficient of load should be determined by the formula (1.15). At $b_{\mathrm{H}} = 18.25$ from (1.15) we obtain that $V_{\mathrm{H}} = 0.06729$. For a number of probability values γ, using formula (3.18), we calculate the corresponding parameters $B(\gamma)$ given in Table 3.2.

With (3.19), it was calculated probabilistically justified safety factors in Table 3.2. They correspond to a combination of Weibull distributions (load) and log-logistic (limit load) variation coefficients of load $V_{\mathrm{H}} = 0.06729$ and limit load $V_{\Pi} = 0.1$. For these variation coefficients according to the formulas (2.3) and (2.7), it was calculated and given safety factors for comparison in Table 3.2. They correspond to the load distributions and limit load according to the normal and log-normal laws. In the case of the considered combination of distributions, the increase in the reliability function at the first load $\gamma \geq 0.999$ leads to higher values of the required safety factor than under normal or log-normal laws.

Considering the m-fold load in the particular case for $b_o = 1$ from (3.11) after transition to the variable $x = \frac{B}{1-G}$, we obtain

$$R_m = B \int_{B}^{\infty} (1 - \exp(B - x))^m \frac{dx}{x^2}. \tag{3.20}$$

We use the binomial expansion of the *m*th power in (3.20)

$$(1 - \exp(B - x))^m = \sum_{i=0}^{m} (-1)^i \binom{i}{m} \exp i(B - x),$$

where $\binom{i}{m} = \frac{m!}{i!(m-i)!}$—binomial coefficients.

As a result of term-by-term integration (3.20) and transformations, we will have an analytic expression for the reliability function for *m* times of the random load of the element:

$$R_m = \sum_{i=1}^{m} (-1)^{i+1} \binom{i}{m} iBe^{iB} E_1(iB). \tag{3.21}$$

To simplify and computerize the calculations when determining the summands of the sum (3.21), if $B \geq 10$ we can use the approximation (3.17) in the form

$$iBe^{iB} E_1(iB) = \frac{(iB)^2 + c_1 iB + c_2}{(iB)^2 + b_1 iB + b_2}, \quad i = 1, 2, \ldots, m. \tag{3.22}$$

The calculations performed with the help of (3.21) showed that similarly to the considered variants, the following conditions are also fulfilled: $R_m > R_1^m$ and $R(m/m_{\Pi}) > R_m$, if $m_{\Pi} > 0$, i.e. it is possible to manage reliability by preloading.

3.2 Reliability Management with the Help of Safety Factors

The considered models of predicting the reliability function under repeated extreme loads assume that the random values of the limit load of the elements during its operation do not change, preserving the initial values, i.e. "fixed in time." This assumption can be justified if the mode of using the element for its intended purpose is relatively stable, and the temperature and other attendant factors are within acceptable limits and practically do not affect the limit load. If there is a random dispersion of the time-fixed limit load, the greatest probability of a sudden failure occurs at the first extreme load, and the conditional probability of failure decreases with each subsequent load. Taking into account this fact, it is advisable to use two values as standardized and predicted indicators of failure-free operation of elements for sudden mechanical failures:

R_1—a reliability function at the first extreme load;
R_m—a reliability function for a given number $m > 1$ of extreme loads.

Each of these indicators has an independent meaning, since the R_1 value determines the level of failure-free operation in the initial (warranty) period of operation, which is especially important to ensure reliability of seasonal use equipment. The indicator R_m characterizes the reliability of the object for a long period of operation (useful lifetime) and allows to predict in advance the possible costs caused by sudden failures.

If the normative values satisfying the consumer are given for each of indicators of failure-free operation: $[R_1]$ and $[R_m]$, then, in designing, the choice of the safety factor $K*$ must ensure the simultaneous fulfillment of two conditions:

$$\begin{aligned} R_1(K^*) &\geq [R_1]; \\ R_m(K^*) &\geq [R_m]. \end{aligned} \tag{3.23}$$

The data analysis on probabilistically justified safety factors (see Tables 2.3, 2.5, 2.7, 3.1 and 3.2) is evidence of the dependence of the required spare quantity on the type of assumed laws of distribution of extreme loads and limit load. Therefore, under conditions of uncertainty regarding the type of load distributions and limit load in practical calculations, it is advisable to use such values of the upper limits for the safety factor, which in the real range of possible variation of the limit load coefficients $V_п$ and the load $V_н$ would provide guarantees of ensuring a given reliability function.

An acceptable practical solütion to the problem of rational reliability management can be obtained on the basis of analysis of calculations of the safety factors conducted using the analysis of various combinations of unimodal distributions for unlimited loads and limit load: normal, log-normal, Weibull, double exponential, power, Sedrakian, logistic, log-logistic, Frechet, as well as combinations of generalized gamma distribution with the Weibull and Frechet laws. By ranking by the size and selection of the largest values of the safety factors, it was defined the upper limits for the safety factor at the first loading, given in Table 3.3. In practice, we usually have to be limited by variation coefficients as the characteristics of dispersion of the

Table 3.3 Upper bounds of probabilistically justified safety factorsby average at the first load

			The reliability function at the first load γ		
$V_н$	ρ	$V_п$	0.99	0.999	0.9999
0.08	16.81	0	1.267	1.453	1.667
		0.1	1.470	1.789	2.169
0.1	13.62	0	1.340	1.586	1.878
		0.1	1.513	1.863	2.276
0.12	11.50	0	1.410	1.724	2.106
		0.1	1.564	1.959	2.433
0.2	7.26	0	1.712	2.351	3.230
		0.1	1.820	2.511	3.425
0.3	5.18	0	2.101	3.278	5.111
		0.1	2.190	3.421	5.323

limit load and load used in engineering calculations. It is no longer practical to consider distribution laws with more than two independent parameters. When choosing a type of suitable distribution laws, it is also necessary to take into account the real upper bounds of variation ranges for variation coefficients of: the limit load—$V_{\text{п}} \leq 0.1$; loads—$V_{\text{н}} \leq 0.3$.

From the combination of the considered combinations of unimodal two-parameter distribution laws for unbounded random variables it was revealed that the largest values of the safety factors for the average $\overline{K}(\gamma)$, corresponding to the high level of the reliability function $\gamma \geq 0.99$, are obtained if the random limit load of an element distributed according to the Weibull law is given, and the random extreme load –distribution of Frechet. This combination of types of distributions of the limit load and load is unfavorable and can later be recommended for use in estimating the upper bounds of probabilistically justified safety factors. An indirect confirmation of this can be the fact that both these distributions belong to the limiting category, and the Weibull law is the limiting distribution of minima, while the Frechet law describes the distribution of the maxima of the random variables [50]. Using the method of single distributions in this case, an expression was obtained for the reliability function of an element under repeated extreme load depending on the safety factor

$$R_m(\overline{K}) = \int_0^1 \exp\left\{-m\left(\frac{\Gamma(1+1/b)}{\overline{K}\Gamma(1+1/\rho)}\right)^{\rho}[-\ln(1-x)]^{-\rho/b}\right\}dx, \tag{3.24}$$

where the values of the parameters of the form b and p of the Weibull and Frechet distributions corresponding to the given values of the load variation coefficients $V_{\text{н}}$ and the load capacity $V_{\text{п}}$ can be found using Table 1.1. With the help of (3.24), solving the equation $R_1(\overline{K}_{1\gamma}) = \gamma$ at $m = 1$, it was determined the values of the upper limits of the safety factor at the first loading $\overline{K}_{1\gamma}$, corresponding to the variation coefficient of the limit load $V_{\text{п}} = 0.1$ $(b = 12.15)$. These values for a number of values of γ are given in Table 3.3.

Such an approach to establishing the relationship between the reliability function and safety factor in managing reliability can compensate, to some extent, the main drawback of the probabilistic “load–limit load” model, which consists in the need to make any assumptions about the type of theoretical distribution laws used.

It should be noted that even with significant amounts of statistical data on extreme loads and limit load of elements, the known statistical consensus criteria [51] do not give an unambiguous answer about the “true” type of theoretical distribution. Therefore, the use of hypotheses adequate to the previous experience on this point is inevitable in engineering practice.

In Table 3.3 In addition to the values $\overline{K}_{1\gamma}$, corresponding to the random load capacity with the variation coefficient $V_{\text{п}} = 0.1$, the upper limits for the safety factors in the case of deterministic limit load are also given (for $V_{\text{п}} = 0$). These boundaries were calculated at $m = 1$ using the expression obtained under the assumption that the load has a Frechet distribution

$$\overline{K}_m(\gamma) = \frac{(1 - 1/_\rho)}{\Gamma(2 - 1/_\rho)} \left(\frac{m}{\ln 1/_\gamma} \right)^{1/_\rho}. \tag{3.25}$$

Of all the distributions considered, this law gives the largest values of the safety factors in the given range of the values of V_H and γ, at $V_\Pi = 0$ (see Table 1.3). The variant equivalent to the deterministic limit load can be practically realized if complete control is provided when elements are manufactured, using preloading with a given constant value of the load P_{max}, and the safety factor should be calculated as a ratio of $\overline{K} = P_{max}/\overline{P}_H$. If we provide practically deterministic and identical limit load for the elements making up the system, then when they are randomly loaded together, the system's reliability function will not depend on its structure and number of elements, but will be determined by the reliability function of one element. In this case, the data shown in Table 3.3, can be used to ensure the necessary level of reliability of such systems.

The use of upper bounds for safety factors, each of which corresponds to a certain level of reliability function, ensures that if the load and limit load are distributed according to any of the listed laws, the safety factor adopted according to the data of Table 3.3, will ensure the reliability function R_1 of an element not lower than the specified value. The upper bound for the safety factor of an element in the case of a given number m of extreme loads having a Frechet distribution can be determined from

$$\overline{K}_m(\gamma) = \overline{K}_{1\gamma} m^{1/_\rho}. \tag{3.26}$$

Using formula (3.26) and the data of Table 3.3 in semilogarithmic coordinates, graphs of the upper limits of the safety factor are plotted as a function of the number of extreme loads m for which the logarithmic scale is used along the abscissa axis. The graphs are shown in Fig. 3.1 and correspond to the reliability function $\gamma = 0.999$.

The Analysis of graphs shows that the most significant effect on the rate of growth of the boundaries for the safety factor with increasing m is the magnitude of the load variation coefficient V_H. A significant decrease in the safety factor due to a decrease in the dispersion of the limit load (V_Π decrease) is possible only for sufficiently small values of V_H. It can also be noted that if the variation coefficient of the extreme load $V_H \geq 0{,}2$, then with the expected number of extreme loads $m > 10$, the guaranteed provision of a sufficiently high reliability function becomes problematic only due to the management of the safety factors.

Summarizing the above information, it can be stated that, based on the application of the probabilistic model of reliability of the "load-limit load" type, the values of the upper bounds of the safety factor can be obtained numerically. Their use in managing reliability provides practical guarantees of ensuring a given reliability function in

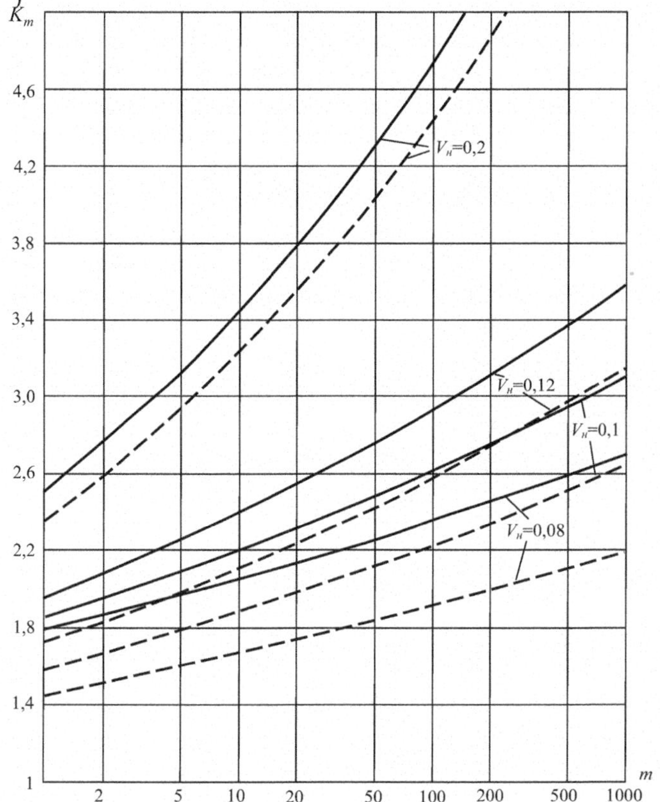

Fig. 3.1 Upper bounds of the safety factor when the element is repeatedly loaded: solid line—when $V_{\text{п}} = 0.1$; dashed line—when $V_{\text{п}} = 0$

case of uncertainty regarding the type of load distributions and limit load. The results obtained allow us to assign safety factors for sudden failures when designing, taking into account the values of variation coefficients of load and limit load, as well as the possible number of extreme loads.

Under repeated loads, in order to ensure the second of the conditions of failure-free operation (3.23) to fulfill, taking into account the magnitude of the variation coefficients $V_{\text{п}}$ and $V_{\text{н}}$, the value of the reliability function, corresponding to a given number of loads, must be determined. In the case of uncertainty regarding the form of the laws of the distribution of the limit load of an element and the load when managing reliability, one should use the expressions (3.24), (3.26) and the data given in Table 3.3.

Let assume the load variation coefficients $V_{\text{н}} = 0.12$ and the limit load $V_{\text{п}} = 0.10$ are known and the normative values of the reliability function at the first loading $[R_1] = 0.999$ and at $m = 100$ loads $[R_{100}] = 0.99$ are given. Then at $\gamma = 0.99$ according to Table 3.3 we find $\overline{K}_{1\gamma} = 1.564$, and then using the formula (3.26) at

$\rho = 11.49$, we determine $\overline{K}_{100}(0.99) = 2.335$. If we take this value for the safety factor $K^* = 2335$ in design, then R_{100} $(2.335) = 0.99$ and the second of conditions (3.23) will be fulfilled. Thus, it follows from the data of Table 3.3 that R_1 $(2.335) > 0.999$ and both conditions of faultness normalization (3.23) will be satisfied. Calculation with the help of numerical integration (3.24) at $m = 1$ and $K^* = 2.335$ allows us to refine the value of R_1 $(2.335) = 0.999845$.

In the cases where the limit load depends on the operating time and can be schematized as a stationary random process, it is permissible to predict the reliability function with the m-fold load using the geometric distribution (1.26). Then the relationship between the reliability function at the first R_1 and m under loads has the form: $R_m = R_1^m$. The upper bound of the probabilisticaly justified safety factor at m loads in this case can be determined, as during the first loading, but for the probability $\gamma^* = \gamma^{1/m}$, i.e. from the equation: $R_1\left(\overline{K}(\gamma)\right) = \gamma^{1/m}$. The safety factor $\overline{K}(\gamma^{1/m})$ obtained in this way will be higher than the upper limit $\overline{K}(m, \gamma)$ corresponding to the time-fixed random limit load scheme.

3.3 Prediction and Management of Reliability Under the Poisson Flow of Loads

Since extreme loads relate to rare random events [9], it is advisable to predict the reliability indexes depending on the operating time assuming that the random extreme load flow is Poisson, in which the distribution density of the load number is determined by the expression (1.43). Then the reliability function can be determined using an expression of the form (1.38), in which, if the limit load is randomly dispersed, conditional probabilities R_i are determined from expressions of the form (2.10) or (2.19), i.e.

$$R_i = \int_0^\infty F^i(P) g(P) dP = \int_0^1 F_1^i(G) dG = \int_0^1 F^i g_1(F) dF. \tag{3.27}$$

Substituting (1.43) and (3.27) into (1.38), we obtain:

$$R(t) = \int_0^\infty g(P) e^{-\overline{m}(t)} \sum_{i=0}^\infty \frac{(\overline{m}(t))^i F^i(P)}{i!} dP. \tag{3.28}$$

Following [32], we can replace the sum of the series in (3.28) with an exponential and obtain the general form of the expression for the reliability function depending on the operating time:

$$R(t) = e^{-\overline{m}(t)} \int_0^\infty e^{\overline{m}(t)F(P)} g(P)dP. \tag{3.29}$$

Passing to (3.29) with unit distributions in (3.27), we obtain two simpler expressions for the reliability function:

$$R(t) = \int_0^1 e^{-\overline{m}(t)[1-F_1(G)]} dG = \int_0^1 e^{-\overline{m}(t)(1-F)} g_1(F)dF. \tag{3.30}$$

If the flow of extreme loads is stationary with a constant intensity $\omega_o = {}^1\!/_{T_o}$, then based on (3.30), the reliability function is determined by the expressions

$$R(t) = \int_0^1 e^{-\omega_o t[1-F_1(G)]} dG = \int_0^1 e^{-\omega_o(t)(1-F)} g_1(F)dF. \tag{3.31}$$

The function of the unit load distribution $F_1(G)$ of type (2.15), which enters into (3.31), depends both on the type of load distributions and on the limit load, and on the size of their parameters. For example, in the case of an unfavorable (in the sense of reliability) combination of Frechet distributions for load and Weibull distributions for the limit load, we have the following expression for this function:

$$F_1(G) = \exp\left\{ -\left(\frac{\Gamma(1 + {}^1\!/_b)}{\overline{K}\Gamma(1 + {}^1\!/_\rho)} \right)^\rho [-\ln(1 - G)]^{-\rho/b} \right\}, \tag{3.32}$$

where ρ and b are parameters of the laws of Frechet and Weibull, uniquely dependent on the variation coefficients of load V_{H} and limit load V_{Π}.

Equation (3.32) also includes a safety factor $\overline{K}$, which basically can provide the necessary level of reliability in the design.

The failure rate at a constant intensity of loads ω_o is determined by the formula:

$$\lambda(t) = \omega_o \left(1 - \frac{\int_0^1 F_1(G) e^{\omega_o t F_1(G)} dG}{\int_0^1 e^{\omega_o t F_1(G)} dG} \right). \tag{3.33}$$

It follows from (3.33) that for a constant loading intensity, the failure rate $\lambda(t)$ is a monotonically decreasing run time function. In the simplest form, when the load and limit load have Weibull or Frechet distributions with the same variation coefficients, and the safety factor $\overline{K} = 1$, we get that $F_1(G) = G$.

It follows from (3.31) an analytical expression for the reliability function as a function of the operating time:

$$R(t) = \frac{1 - e^{-\omega_o t}}{\omega_o t}. \tag{3.34}$$

Proceeding to the dimensionless variable $\tau = \omega_o t$, you can get an expression for the average operating time before a sudden failure

$$T = \int_0^\infty R(t)dt = \frac{1}{\omega_o} \int_0^\infty \frac{1 - e^{-\tau}}{\tau} d\tau. \tag{3.35}$$

Using the tabular [46] integral

$$\int_0^x \frac{1 - e^{-\tau}}{\tau} d\tau = E_1(x) + \ln \mid x \mid + C_o,$$

where $E_1(x)$ is the integral exponential function;
$C_o = 0.57721\ldots$—Euler constant;

allows us to conclude that the integral in (3.35) is divergent and, therefore, within the framework of the model under consideration, there is no finite value for the mean time to failure. Such a result is obtained when $\overline{K} = 1$, but taking into account that with the increase of the safety factor, $R(t)$ increases. It is logical to expect that at $\overline{K} > 1$, the value is $T \to \infty$. Therefore, for a stationary Poisson flow of loads and a random unlimited over-scattering of the limit load, some of the loaded elements will remain operable for any increase in operating time.

In the special case, if extreme loads and limit load are similar random variables and distributed according to the Weibull law with the same parameter of form b, then we can obtain an analytical expression from (3.31) to predict the reliability function, depending on the operating time. In this case, the function of the unit load distribution $F_1(G) = 1 - (1 - G)^{\overline{K}^b}$, and after substituting this into (3.31), we obtain

$$R(t) = \int_0^1 e^{-\omega_o t (1-G)^{\overline{K}^b}} dG. \tag{3.36}$$

The replacement of the variable $x = (1 - G)^{\overline{K}^b}$ leads this integral to the form

$$R(t) = \frac{1}{\overline{K}^b} \int_0^1 e^{-\omega_o t x} x^{\frac{1}{\overline{K}^b} - 1} dx. \tag{3.37}$$

After integrating and transforming (3.37), it is expressed with tabulated gamma functions [46] in the form

$$R(t) = \frac{\Gamma\left(1 + {}^{1}\!/\!_{\overline{K}^{b}}\right) - \Gamma\left(1 + {}^{1}\!/\!_{\overline{K}^{b}},\ \omega_{o}t\right)}{(\omega_{o}t)^{1/\overline{K}^{b}}} + e^{-\omega_{o}t}, \tag{3.38}$$

where $\Gamma(\alpha, \omega_{o}t) = \int\limits_{\omega_{o}t}^{\infty} e^{-y}y^{\alpha-1}dy$ is an incomplete gamma function.

It can be shown that when $\overline{K} = 1$, (3.38) coincides with (3.34).

The calculation of complete $\Gamma\left(1 + {}^{1}\!/\!_{\overline{K}^{b}}\right)$ and incomplete $\Gamma\left(1 + {}^{1}\!/\!_{\overline{K}^{b}},\ \omega_{o}t\right)$ gamma functions in (3.38) can be carried out using the computer mathematical package Mathcad. For approximate calculations of the incomplete gamma function in engineering calculations, we can use [54] its Legendre decomposition into a continuous fraction:

$$\Gamma\left(1 + {}^{1}\!/\!_{\overline{K}^{b}}, \omega_{o}t\right) \equiv \frac{e^{-\omega_{o}t}(\omega_{o}t)^{1+1/\overline{K}^{b}}}{\omega_{o}t - \cfrac{1/\overline{K}^{b}}{1+\cfrac{1}{\omega_{o}t+\cfrac{1-1/\overline{K}^{b}}{1+\cfrac{2}{\omega_{o}t+2-1/\overline{K}^{b}}}}}}. \tag{3.39}$$

For sufficiently large values of the safety factor, if $\omega_{o}t \geq 10$, then the incomplete gamma function is $\Gamma\left(1 + {}^{1}\!/\!_{\overline{K}^{b}},\ \omega_{o}t\right) \approx 0$ and $e^{-\omega_{o}t} \approx 0$. In this case, prediction of the reliability function can be performed using an approximate formula:

$$R(t) = \frac{\Gamma\left(1 + {}^{1}\!/\!_{\overline{K}^{b}}\right)}{(\omega_{o}t)^{1/\overline{K}^{b}}}. \tag{3.40}$$

On the assumption of (3.40) and assuming that $\Gamma\left(1 + /1_{\overline{K}^{b}}\right) \approx 1$, for a probabilistically justified safety factor with a sufficiently large guaranteed probability $R(t_{\gamma}) = \gamma > 0.95$, we obtain an approximate dependence:

$$\overline{K}_{\gamma}(t_{\gamma}) = \left(\frac{\ln \omega_{o}t_{\gamma}}{\ln {}^{1}\!/\!_{\gamma}}\right)^{1/b}, \tag{3.41}$$

which gives an obviously overestimated assessment for $\overline{K}_{\gamma}(t_{\gamma})$ and is valid for $\omega_{o}t_{\gamma} \geq 10$.

To obtain an analytical expression for the reliability function, on the assumption of (3.30), it is also possible in the case of similar load and limit load if they have Frechet distribution. Then, in accordance with (2.59) and (3.30), we have

$$R(t) = \int_0^1 e^{-\overline{m}(t)\left(1 - G^{\frac{1}{\overline{K}^\rho}}\right)} dG. \tag{3.42}$$

Changing a variable $x = G^{1/\overline{K}^\rho}$ in (3.42), we obtain an expression convenient for integration

$$R(t) = e^{-\overline{m}(t))}\overline{K}^\rho \int_0^1 e^{\overline{m}(t)x} x^{\overline{K}^\rho - 1} dx. \tag{3.43}$$

The integral in (3.43) is tabular [21] and is computed in closed form for integer values $\overline{K}^\rho$. The result is an expression for the reliability function in the form of a finite sum:

$$R(t) = (k+1) \times \left[\frac{1}{\overline{m}(t)} + \sum_{i=1}^{k} (-1)^i \frac{k(k-1)\ldots(k-i+1)}{[\overline{m}(t)]^{i+1}} + (-1)^{k+1} \frac{k! e^{-m(t)}}{[\overline{m}(t)]^{k+1}} \right], \tag{3.44}$$

where $k = \overline{K}^\rho - 1$.

It is convenient to use formula (3.44) only for sufficiently small integer values $\overline{K}^\rho$. Therefore, in the general case it is advisable to use an expression of the form (3.42), using numerical integration. Based on (3.33), $\overline{K} = 1$, and $F_1(G) = G$, after integration, an expression for the failure rate can be obtained in an explicit form:

$$\lambda(t) = \frac{e^{\omega_o t} - \omega_o t - 1}{t(e^{\omega_o t} - 1)}. \tag{3.45}$$

It can be shown that in this particular case $\lim_{t \to 0} \lambda(t) = \frac{\omega_o}{2}$, and further with an increase in the operating time, the failure rate function decreases monotonically, tending to zero. In the general case, it follows from (3.33) that the initial (largest) value of the failure rate at $t = 0$ is given by the formula:

$$\lambda(0) = \omega_o \left(1 - \int_0^1 F_1(G) dG \right) = \omega_o (1 - R_1). \tag{3.46}$$

Therefore, in comparison with the intensity of extreme loads ω_o, the initial value of the failure rate decreases in proportion to the probability of failure at the first load. The expression (3.46) coincides with (1.41). However, firstly, the probabilities included in them R_1 and $R_1(K)$ are calculated differently. Secondly, in the case of a random dispersion of the limit load, the failure rate is monotonically decreasing, starting from, the runtime function in difference from $\lambda(0)$ of the value independent of the operating time, calculated by the formula (1.41).

Figure 3.2 shows the graphs of the reliability function as a function of the dimensionless operating time $\tau = \omega_o t$. The curves in Fig. 3.2a are constructed with the help of the general expression (3.31) at $F_1(G) = 1 - (1 - G)^{\overline{K}^b}$, which corresponds to the Weibull distribution for the load and the limit load. At the same time $b = 12.15$ (variation coefficient of the load and limit load $V = 0.1$).

Figure 3.2b presents also graphs the reliability function, which correspond to an unfavorable combination of load distributions according to the Frechet law and the

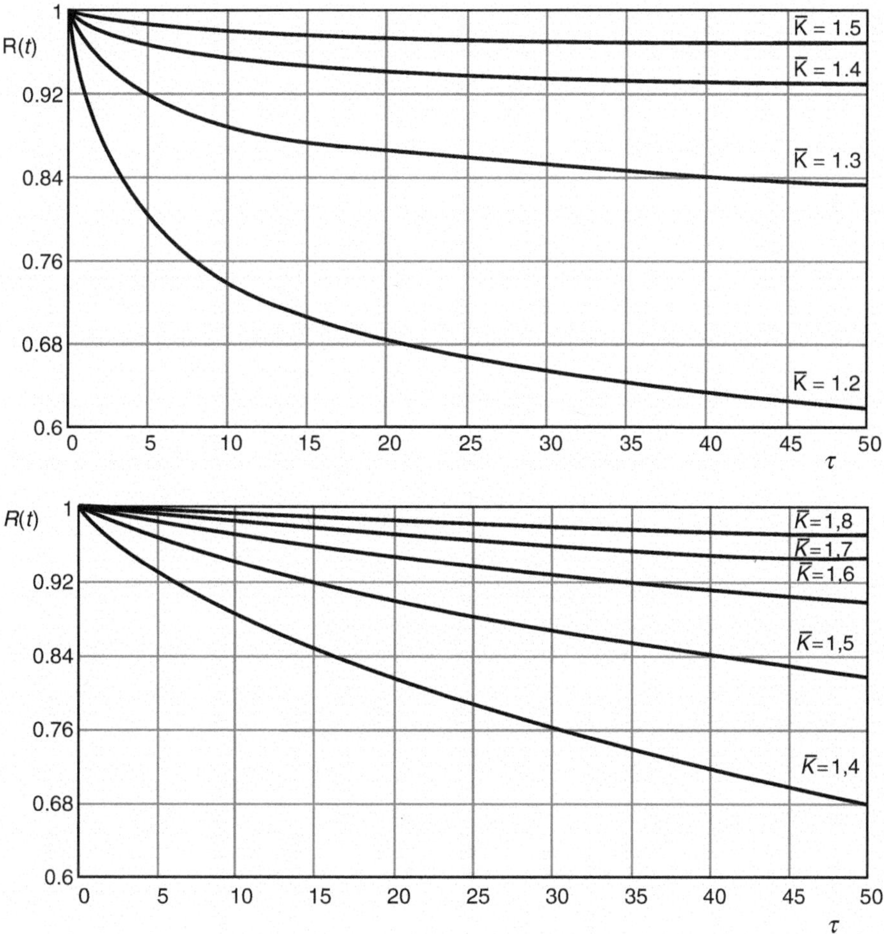

Fig. 3.2 Curve charts of changing the reliability function depending on the operating time and the safety factor at: (**a**) Weibull law for load and limit load; (**b**) a combination of the laws of Frechet and Weibull

limit load according to the Weibull law. These graphs are constructed with the help of the general expression (3.31), in which the function of the unit load distribution was determined from the formula (3.32), and the coefficients of load variation and the limit load $V_{\rm H} = V_{\Pi} = 0.1$ were provided by the values of the shape parameters: $\rho = 13.62$ and $b = 12.15$.

The analysis of the graphs shown in Fig. 3.2a it is obvious a possibility of effective management of reliability by changing the safety factor by the mean values $\overline{K}$. At the same time, it is possible to provide the required reliability function γ, taking into account that the following condition must be fulfilled within the period τ_γ:

$$R(\tau_\gamma) \geq \gamma, \quad \text{when } \tau \leq \tau_\gamma. \tag{3.47}$$

The curve charts in Fig. 3.2 indicate the possibility of stabilizing the level of failure-free operation of the elements due to the increase of safety factors and the corresponding decrease in the influence of the operating time on the reliability function. From a comparison of the charts in Fig. 3.2a, b) constructed at the same level of random dispersion of load and limit load, it follows that the type of load distributions and limit load used has a significant effect on the nature of the change and the reliability function. Therefore, in the context of uncertainty of the information regarding the actual form of these distributions, in order to ensure a guaranteed result in reliability management, we can recommend the use of expression (3.32) for the unit load distribution function $F_1(G)$ included in the general expression (3.31) to predict the reliability function. Table A.3 of the appendix gives data on the guaranteed dimensionless operating time τ_γ, which satisfies the condition (3.47). This condition is ensured with an unfavorable combination of the load distributions and limit load, to which expression (3.32) corresponds. From a comparison of the Table data A.3 with the results of calculations, given in Table A.1 similar in meaning and content, it follows that there is a significant effect of random scattering of the limit load of the elements by the amount of guaranteed operating time τ_γ. This conclusion indicates the expediency and effectiveness of reducing the degree of dispersion of the limit load while ensuring reliability.

Let us consider a numerical example illustrating the nature of the failure rate behavior as a function of the operating time for a stationary Poisson load flow. Let us assume the average period between loads of unit operating time as $T_o = 100$ and, consequently, the intensity of the Poisson load flow $\omega_o = 0.01$ (shown in Fig. 3.3 as a dashed line). If the random load has a Weibull distribution with a variation coefficient of $V_{\rm H} = 0.1$ and a shape parameter of $b_{\rm H} = 12.15$, then with a deterministic constant limit load and a safety factor of $K = 1$, the failure rate in accordance with expression (1.41) is determined (see Table 1.2) as follows:

$$\lambda = 0.01 \cdot \exp\left\{-[\Gamma(1 + {}^1\!/_{12.15})]^{12.15}\right\} = 0.00549.$$

This constant value of the failure rate is shown in Fig. 3.3 as a horizontal line 1. Below Fig. 3.3 is a graph of the failure rate in the case where the limit load is a

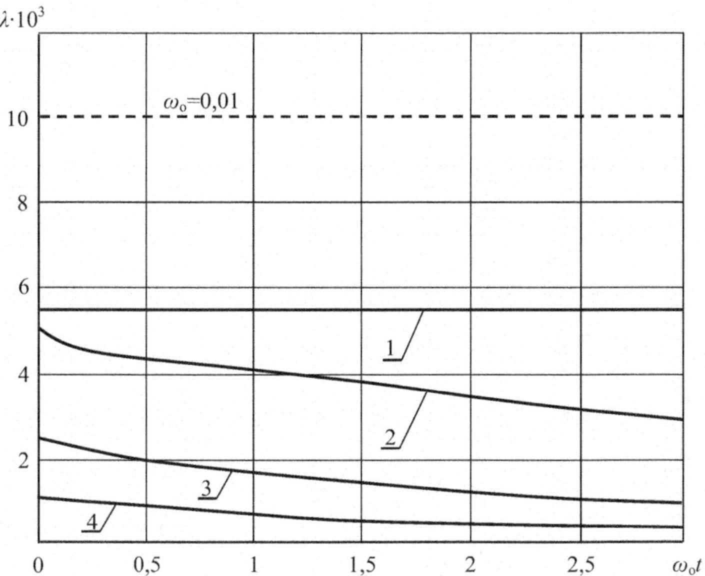

Fig. 3.3 Sudden failure rate function graphs for a stationary Poisson flow of loads

random variable fixed in time, distributed in the same way as the Weibull load with the parameter $b_{\Pi} = b_{H} = 12.15$, and the safety factor $\overline{K} = 1$. This chart (curve 2) is constructed by the formula (3.45) and has a decreasing character depending on the dimensionless operating time $\omega_o t$. If the safety factor $\overline{K} > 1$, then in the case under consideration $F_1(G) = 1 - (1 - G)^{\overline{K}^b}$, the failure rate is calculated using the general formula (3.33) using numerical integration. In this way, Fig. 3.3 builds: curve 3 (for $\overline{K} = 1.1$) and curve 4 (for $\overline{K} = 1.2$). They have a monotonically decreasing character, and also show the effect of increasing the safety factor on reducing the failure rate.

In particular, the initial (largest) value of the failure rate in the case under consideration with allowance for (3.46) can be determined by the formula:

$$\max \lambda(t) = \lambda(0) = \frac{\omega_o}{\overline{K}^b + 1}. \tag{3.48}$$

This expression allows you to designate the size of the safety factor when designing, managing reliability based on the requirements for the level of failure-free operation of the element in the initial period of operation. The decrease in the failure rate function $\lambda(t)$ indicates that, in contrast to the variant, when there is no accidental dissipation of the limit load of the elements, in the presence of such a scattering and a stationary Poisson load flow, the failure flow for the set of elements differs from the Poisson one. In this case, the distribution of the random time between failures will differ from the exponential. The decreasing nature of the

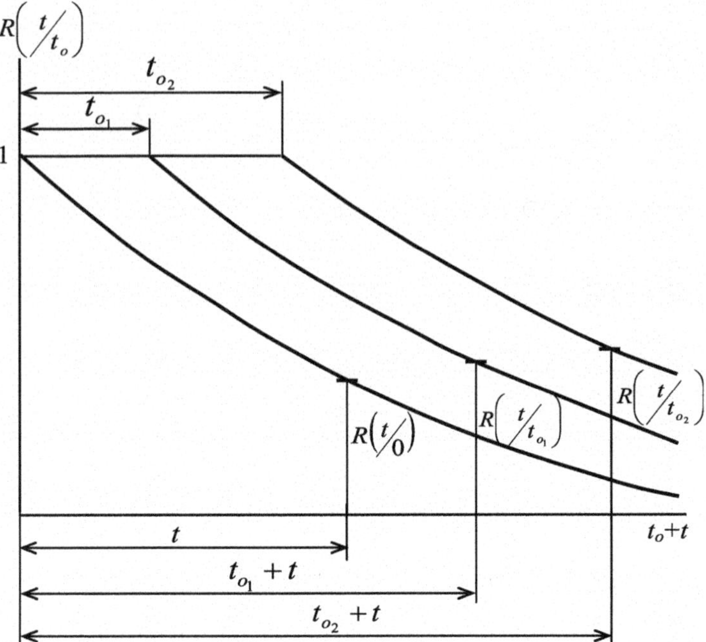

Fig. 3.4 Graphs for changing a conditional reliability function with an increase in the worked-out period

failure rate depending on the service life for brittle fracture of steelworks is confirmed by the statistical data given in [26].

The maximum value of the failure rate, determined by the expression (3.46), can be reduced in comparison with the intensity of loads ω_o due to the corresponding setting of the safety factors. In the case of Weibull distribution, the formula (3.48) is used for such loads and the limit load. If we use the Frechet distribution, then $\max \lambda(t) = \frac{\omega_o}{\overline{K}^{\rho}+1}$. In the case of an unfavorable combination of load distributions and limit load (Frechet and Weibull distributions), reliability management can be performed using data on the upper limits of the safety factor given in Table 3.3. So, if the values of variation coefficients of the load $V_{\text{H}} = 0.2$ and the limit load $V_{\Pi} = 0.1$ are required to reduce the maximum failure rate in comparison with the load intensity by 100 times, then it follows from Table 3.3 that for this purpose the safety factor $\overline{K} = 1.82$, is sufficient, and to decrease max$\lambda(t)$ to $0.001\omega_o$, it is necessary to have $\overline{K} = 2.511$.

The consequence of a monotonous decrease of the intensity of sudden failures will be a feature in the change in the corresponding conditional reliability function. The conditional reliability function $R\left(t/_{t_o}\right)$ depends on two arguments: t_o—a worked-out failure-free preliminary period and t—a subsequent operating time. The expression for determining a conditional reliability function has the form

$$R\left(t/t_o\right) = \frac{R(t_o + t)}{R(t_o)}. \tag{3.49}$$

A typical regularity for the case of a monotonically decreasing failure rate of the behavior of the function $R\left(t/t_o\right)$ is that for the same value of the subsequent operating time t, with an increase in the preliminary period t_o, the conditional reliability function increases [47]. This is illustrated in Fig. 3.4, where, given that $0 < t_{o_1} < t_{o_2}$, the conditional probability values increase: $R(t/0) < R\left(t/t_{o_1}\right) < R\left(t/t_{o_2}\right)$.

Therefore, when designing a basic characteristic that determines the level of reliability for sudden failures, the unconditional reliability function $R(t) = R(t/0)$ is a function that, in the process of operation, should not go down in the rest of working elements. An increase of the conditional reliability function $R\left(t/t_o\right)$ with an increase of the worked out preliminary period t_o may be accompanied by the replacement of some failed (weakest) elements. However, implementation of preventive replacements of exploited elements in the case of sudden mechanical failures, caused only by overloads, has no effect.

Chapter 4
Prediction of System Reliability

Any object in the reliability theory can be considered as an element or as a system consisting of elements that form a certain structure in the sense of reliability. According to the structure of the system where mechanical failures occur, they are divided into consecutive (Fig. 4.1a) and fault-tolerant, including parallel ones (Fig. 4.1b) and those which have a reserve of survivability (Fig. 4.1c). When analyzing systems, it is necessary to distinguish the structure in a constructive-functional sense and in terms of reliability.

A failure of a consistent system in terms of the system reliability occurs when at least one of its elements is rejected. Such systems do not have a reserve of fault tolerance (survivability) and, therefore, the system reliability is determined by the reliability of each of the elements, as well as by their quantity. In a system with a parallel structure, a failure occurs in the event of the failure of all its elements. A parallel structure with a limited total number of elements provides a loaded reserve for the system. In mechanical systems, the parallel structure in pure form is rarely used due to complexity of the constructive implementation and increased costs. Therefore, often in machines instead of a loaded reserve, an unloaded replacement by substitution is actually used, considering failures of the replaced elements for the system to be non-critical, i.e. not causing significant damage and losses.

A generalization of the two variants of the structure of systems is a structure with a loaded reserve of survivability (Fig. 4.1c), when a critical failure occurs in a system consisting of n identical elements, with the simultaneous rejection of any l elements. In practice, systems with a loaded survivability reserve can include cluster, threaded, riveted or pin joints, multi-row chain or belt drives, as well as some multi-element building structures. If $l \ll n$, then the loading conditions of the elements in such a system change little until the onset of a critical failure. In these systems, it is permissible to produce a certain number of replacements (rebuilds) of simultaneously failed elements without the appearance of a security threat, significant material damage or other unacceptable consequences. The property of the system to maintain limited up state in case of failures of its constituent parts is called fault

O. Grynchenko, O. Alfyorov, *Mechanical Reliability*,
https://doi.org/10.1007/978-3-030-41564-8_4

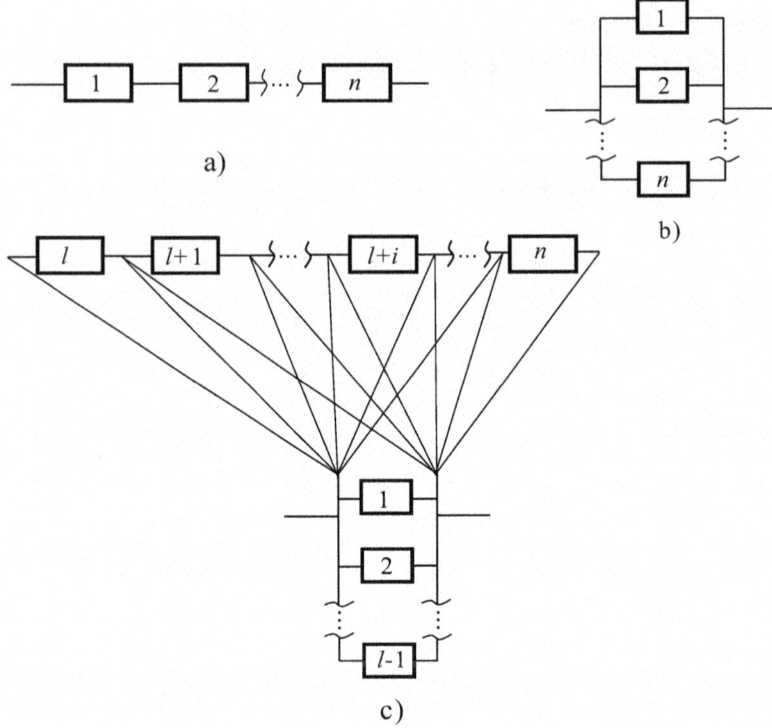

Fig. 4.1 Schemes of structures of multi-element systems: (**a**) consistent; (**b**) parallel; (**c**) with a loaded reserve of survivability

tolerance or survivability [27, 28]. From Fig. 4.1 it is obvious that when $l = 1$, the structure of a multi-element system is consistent, and when $l = n$, it is parallel.

We can consider two schemes of extreme loads of elements in the system: separate (independent) and common (combined) loading. When the elements are independently and randomly loaded in the system, it is assumed that the random sequences of discrete loads (see Fig. 1.1) that act separately on each element do not depend on each other in terms of the magnitude of the extreme loads: P_{H_1}, $P_{\mathrm{H}_2}, \ldots, P_{\mathrm{H}_m}$, and the moments of loading: $t_1, t_2, \ldots, t_m$. Therefore, sudden failures of elements in the system should be considered as independent random events. This assumption usually substantially simplifies the calculation formulas for the reliability function of systems with different structures.

However, the main source of extreme loads acting simultaneously on the elements of one system is often some common load factors that, when transformed, are distributed among the elements in a non-random way. This can be, for example, the torque on the drive shaft of the gearbox, distributed over its elements, or the tractive force of the tractor distributed over the working parts of the tillage unit. In these cases, the condition of independence of extreme loads in one system is violated and a scheme of general or combined loading can be used. Under such a scheme, the extreme load acting simultaneously on each element of the system is defined as the product: $P_{Hi} = \theta_i P_H$, where P_H is a general random load component in the system; θ_i—a parameter that determines the load distribution across the elements of the system and the degree of dependence between the elements on the general loading. We call common or combined loading a simultaneous extreme loading of all elements in one system if one can neglect the random dispersion of the quantities θ_i and regard them as constant proportionality coefficients between the common random load component P_H and the load P_{Hi}, acting on the ith element.

The scheme of combined loading of elements is more applicable when analyzing the reliability of individual units or machine assemblies, for example, a multi leaf suspension of a vehicle. To a lesser extent, it corresponds to reality if a larger system is considered, for example, the transmission or the running system of the machine as a whole. The prediction of the reliability of a complex object as a system must begin with identification of those subsystems within which extreme loads of elements can be considered as combined, and loads of such subsystems among themselves is then treated as separately independent.

4.1 Systems with Serial Structure

In machines, the most common structure in terms of reliability is serial (Fig. 4.1a), when it is considered that the failure of any of the n elements results in the failure of the entire system. When analyzing the reliability of systems, it is necessary to take into account the form of the loading diagram, as well as the presence or absence of random scattering of the limit load of elements.

The distribution function of the limit load of a serial system consisting of n independent in terms of the limit load of elements is given by the following expression.

$$G_c(P_\Pi) = 1 - \prod_{i=1}^{n} (1 - G_i(P_{\Pi i})), \tag{4.1}$$

where $G_i(P_{\Pi i})$ is a distribution functions of the random limit load of the elements of the system.

Then, as a result of differentiating (4.1), we obtain that.

$$dG_c(P_\Pi) = g_c(P_\Pi)dP_\Pi = -d\left(\prod_{i=1}^{n} (1 - G_i(P_{\Pi i}))\right), \tag{4.2}$$

where $g_c(P_\Pi)$ is a density distribution of the limit load of a serial system.

Let us consider a case of the common (combined) loading of elements in the system in the presence of a random scattering of their limit load. In accordance with the flow diagram of stochastically independent discrete random loads (see Fig. 1.1b), the reliability function of the system after exposure to it of extreme loads can be determined from the expression:

$$R_c(m) = \int_o^{\infty} F^m(P) g_c(P) dP, \tag{4.3}$$

where *F*(*P*)—a distribution function of the total load on the system.

Taking into account (4.2) and (4.3), that the reliability function of a serial system consisting of n jointly loaded elements can be determined by means of the dependence:

$$R_c(m) = -\int_o^{\infty} [F(P)]^m d\left(\prod_{i=1}^{n}(1 - G_i(P))\right). \tag{4.4}$$

If extreme loads on the system P_{H} and the limit load $P_{\Pi i}$ of the elements have Weibull distributions with distribution functions.

$$\begin{aligned} F(P_{\text{H}}) &= 1 - \exp\left[-\left(\frac{P_{\text{H}}}{a_{\text{H}}}\right)^b\right]; \\ G_i(P_{\Pi i}) &= 1 - \exp\left[-\left(\frac{\theta_i P_{\Pi i}}{a_{\Pi i}}\right)^b\right]; \quad i = 1, 2, \ldots, n, \end{aligned} \tag{4.5}$$

then the expression (4.4) can be reduced by means of elementary transformations to the form:

$$R_c(m) = -\int_o^{\infty}\left\{1 - \exp\left[-\left(\frac{P}{a_{\text{H}}}\right)^b\right]\right\}^m d\left(1 - \exp\left[-\left(\frac{P}{a_{\Pi c}}\right)^b\right]\right), \tag{4.6}$$

where

$$a_{\Pi c} = \left(\sum_{i=1}^{n}\left(\frac{\theta_i}{a_{\Pi i}}\right)^b\right)^{-1/b}. \tag{4.7}$$

Consequently, the limit load of the system is also distributed according to the Weibull law, but with the scale parameter determined by formula (4.7), and with the same parameter of form b as in the elements.

Making a change of a variable in (4.6)

$$y = 1 - \exp\left[-\left(\frac{P}{a_{\mathrm{H}}}\right)^{b}\right] \tag{4.8}$$

and taking into account that

$$\exp\left[-\left(\frac{P}{a_{\Pi c}}\right)^{b}\right] = (1-y)^{\left(\frac{a_{\mathrm{H}}}{a_{\Pi c}}\right)^{b}},$$

we reduce the integral (4.6) to the form

$$R_c(m) = \left(\frac{a_{\mathrm{H}}}{a_{\Pi c}}\right)^{b} \int_0^1 y^m (1-y)^{\left(\frac{a_{\mathrm{H}}}{a_{\Pi c}}\right)^{b}-1} dy. \tag{4.9}$$

The integral in (4.9) is a complete beta function, therefore

$$R_c(m) = \left(\frac{a_{\mathrm{H}}}{a_{\Pi c}}\right)^{b} B\left(m+1; \left(\frac{a_{\mathrm{H}}}{a_{\Pi c}}\right)^{b}\right). \tag{4.10}$$

Expressing the beta function with the help of gamma functions and taking into account (4.7), we obtain [14, 20] the calculated formula for predicting the reliability function of a serial system in the form

$$R_c(m) = \frac{\Gamma(1+\Omega)\Gamma(1+m)}{\Gamma(1+\Omega+m)}, \tag{4.11}$$

where $\Omega = \frac{\chi}{\overline{K}_{\min}^{b}}$; $\chi = \sum_{i=1}^{n}\left(\frac{\overline{K}_{\min}}{\overline{K}_i}\right)^{b}$; $\overline{K}_i = \frac{\overline{P}_{\Pi i}}{\theta_i \overline{P}_{\mathrm{H}}}$—safety factors by the means: limit load $\overline{P}_{\Pi i}$ and extreme loads $\overline{P}_{\mathrm{H}i} = \theta_i \overline{P}_{\mathrm{H}}$ on the elements; $\overline{K}_{\min}$ the safety factor for the most loaded element of the system.

The value χ that is within the limits $1 \leq \chi \leq n$ should be treated as a conditional number of elements in the system, reduced to the most loaded one. The parameter of the form b of distributions (4.5) is determined only by the value of the overall variation coefficient of V of the extreme loads and the limit load of the elements. Table A.4 in the Appendix gives some values of the function (4.11) for non-integral (average) values of the total number of extreme loads m.

From (4.11) for integer values of the total number of loads m, we can obtain an expression for the probability of a system reliability function as a function of the number of loads in the form analogous to (2.34):

$$R_c(m) = \prod_{j=1}^{m} \frac{j\overline{K}_{\min}^{b}}{j\overline{K}_{\min}^{b} + \chi}. \tag{4.12}$$

The analysis of the structure of this formula shows that each subsequent load in the product (4.12) corresponds a probabilistic multiplier in meaning, which is greater than the previous ones. This is the fundamental difference between the model (4.12) and the currently recommended [9] models of mechanical reliability, in the construction of which it is assumed that the reliability function with each successive loads does not change, and then $R_c(m) = R_c^m(1)$. This understatement decreasing $R_c(m)$ introduces an error that is related to ignoring the stability of the limit load of multiply loaded elements, as well as the unevenness of "weeding out" of the "weakest" elements in the initial and subsequent loading.

With the help of (4.12), one can obtain an expression similar to (2.35) for the conditional reliability function $R_c\left({}^{m}/_{m_\Pi}\right)$, which determines such a probability for a serial system that has already withstood without failures the first m_Π preliminary loads:

$$R_c\left({}^{m}/_{m_\Pi}\right) = \frac{R_c(m_\Pi + m)}{R_c(m_\Pi)} = \prod_{j=m_\Pi+1}^{m_\Pi+m} \frac{jK_{\min}^{b}}{jK_{\min}^{b} + \chi}. \tag{4.13}$$

As follows from (4.13), an increase in the number of preloads m_Π with the conservation of m increases the conditional reliability function of the system $R_c\left({}^{m}/_{m_\Pi}\right)$. This is a theoretical justification to use the method of managing reliability by applying a "power" run-in: performing the required number of control loads before beginning service. Modes of conducting such a run-in can be chosen on the basis of (4.13). Fig. 4.2 shows how an increase in the number of preloads m_Π can affect the increase in the reliability function of a group of the threaded joint consisting of six equally loaded bolts. In this case, the same stock factors for the elements $\overline{K}_i = 1.7$ are adopted, and the coefficient of load variation and limit load is assumed to be V = 0.1.

By analogy with expression (2.42), based on (4.12), an approximate formula can be obtained that determines the probabilistically justified safety factor for the most loaded element of a serial system for a sufficiently large ($m > 5$) number of combined extreme loads:

$$\overline{K}_{\min}(m, \gamma) = \left\{ \frac{\chi[C_o + \ln(m + 0.51)]}{\ln {}^{1}/_{\gamma}} \right\}^{1/b}, \tag{4.14}$$

where $C_o = 0.57721\ldots$—Euler constant.

When determining a value of the safety factor $\overline{K}_{\min}(m, \gamma)$, which provides a predetermined reliability function γ in a serial system after m-fold load, it is necessary to preset values of the relations $\alpha_i = \frac{\overline{K}_{\min}}{\overline{K}_i}$, $i = 1, 2, \ldots, n$ and in advance calculate the reduced number of elements in the system $\chi = \sum_{i=1}^{n} \alpha_i$. Then, according

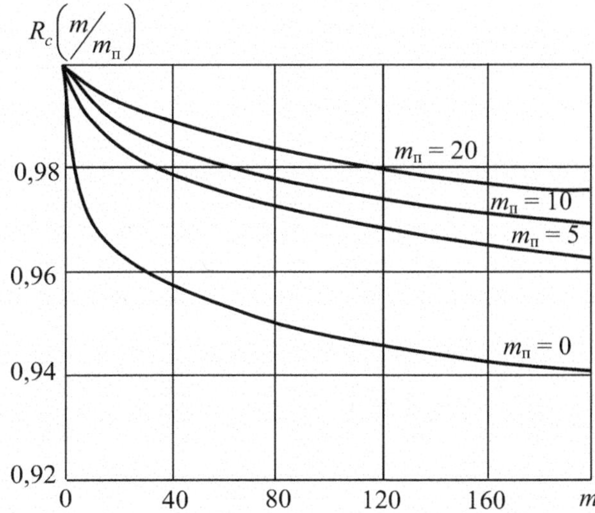

Fig. 4.2 The effect of the number of preloads $m_П$ to increase the reliability of the threaded joint

to the formula (4.14), the probabilistically justified safety factor $\overline{K}_{\min}(m,\gamma)$ for the most loaded element and the safety factors for the remaining elements are determined: $\overline{K}_i = \frac{\overline{K}_{\min}(m,\gamma)}{\alpha_i}$. The ensemble of safety factors $\overline{K}_i$, $i = 1, 2, \ldots, n$ obtained in this way will ensure the level of system faultness when it is being designed.

If in a serial system all elements must be identical in terms of reliability: $\overline{K}_i = \overline{K}$ and, accordingly $\chi = n$, the probabilistically justified safety factors for all elements of the system are determined by the formula:

$$\overline{K} = \overline{K}(m,\ \gamma) \cdot n^{1/b}, \tag{4.15}$$

in which the safety factor $\overline{K}(m,\gamma)$ is determined from the expression (2.42) for one element.

In many cases the costs of ensuring the mechanical reliability of systems from sudden failures can be approximately estimated as proportional to the sum or the average value of the safety factors for the elements: $\overline{\overline{K}} = \frac{1}{n}\sum\limits_{i=1}^{n}\overline{K}_i$. Therefore, the analysis of reliability in the design it makes sense to conduct using dependencies that include an average of $\overline{\overline{K}}$ safety factors for the elements of a serial system. Analogous to the formula (4.12), one can obtain an expression to calculate the reliability function of a serial system under simultaneous multiple loads of elements in the form.

$$R_c(m) = \prod_{j=1}^{m} \frac{j\overline{\overline{K}}^b}{j\overline{\overline{K}}^b + \overline{\chi}}, \tag{4.16}$$

where $\overline{\chi} = \sum_{i=1}^{n} \left(\frac{\overline{\overline{K}}}{\overline{K}_i}\right)^b$ a number of elements in the system, reduced to a conditional element with an average safety factor.

Let's consider and convert the value:

$$\left(\frac{\overline{\chi}}{n}\right)^{1/b} = \overline{\overline{K}}\left(\frac{1}{n}\sum_{i=1}^{n}\frac{1}{\overline{K}_i^b}\right)^{1/b} = \left(\frac{1}{n}\sum_{i=1}^{n}\frac{1}{\overline{K}_i^b}\right)^{1/b} \Big/ \frac{n}{\sum_{i=1}^{n}\frac{1}{\overline{K}_i}}. \tag{4.17}$$

In the numerator of expression (4.17) we have a generalized power mean of the values $1/\overline{K}_i$, and there is a harmonic mean of these same quantities in the denominator, which is always less than the power mean for the denominator at $b > 0$. Equality between the averages can only be if all $\overline{K}_i$ are the same. Therefore, it is always $\left(\frac{\overline{\chi}}{n}\right)^{1/b} \geq 1$ and, consequently $\overline{\chi} \geq n$. Based on (4.16), the foregoing allows to conclude that when the condition is satisfied $\overline{\overline{K}} = const$, we have an upper bound for the reliability function of the serial system:

$$R_c(m) \leq \prod_{j=1}^{m} \frac{j\overline{\overline{K}}^b}{j\overline{\overline{K}}^b + n}. \tag{4.18}$$

It is obvious that $\max R_c(m) = \prod_{j=1}^{m} \frac{j\overline{\overline{K}}^b}{j\overline{\overline{K}}^b + n}$ is only achieved if all the safety factors for the elements $\overline{K}_i$ are the same and $\overline{K}_i = \overline{\overline{K}}$. Therefore, to provide a reliable version of building a serial system, it is rational to perform it from the equally loaded elements, i.e. elements with the same or close to the safety factors. It is confirmed that in the design of serial systems optimal by the criterion of mechanical reliability, instead of the known principle of "strength balance", it is advisable to use a similar in meaning, but more general in content, principle of equal reliability of the elements entering the system.

In the case of a stationary Poisson flow of extreme loads acting jointly and simultaneously on the elements of a serial system with load distribution functions and bearing capacities of the form (4.5) with the help of the general expression (3.36), we obtain a dependence of the reliability function of the serial system on the operating time in the form:

$$R_c(t) = \int_0^1 e^{-\omega_o t(1-G)^{K_c^b}} dG, \tag{4.19}$$

where $K_c = a_{\text{H}}^{-1} \cdot \left(\sum_{i=1}^{n}\left(\frac{\theta_i}{a_{\Pi i}}\right)^b\right)^{-1/b}$.

When predicting the reliability function of the system $R_c(t)$, one can also use the expressions (3.38) and (3.40), substituting into K_c instead of $\overline{K}$. Using formula (3.41), we can approximately estimate the value of the probabilistically justified safety factor $K_c(t_\gamma)$ of the system.

Considering a system of n equally reliable elements with safety factors $\overline{K}$, we obtain for the reliability function n the expression from (4.19):

$$R_c(t,n) = \int_0^1 e^{-\omega_o t (1-G)^{\frac{\overline{K}^b}{n}}} dG. \tag{4.20}$$

Proceeding from (4.20), by analogy with (3.36) and (3.38), one can obtain an expression for predicting the reliability function of a serial system under simultaneous loading in the analytical form:

$$R_c(t,n) = \frac{\Gamma\left(1 + {}^{n}\!/\!_{\overline{K}^b}\right) - \Gamma\left(1 + {}^{n}\!/\!_{\overline{K}^b}, \omega_o t\right)}{(\omega_o t)^{n/\overline{K}^b}} + e^{-\omega_o t}. \tag{4.21}$$

Analogously to (3.40) at $\omega_o t \geq 10$, we can use the approximate formula

$$R_c(t,n) = \Gamma\left(\frac{\left(1 + {}^{n}\!/\!_{\overline{K}^b}\right)}{(\omega_o t)^{n/\overline{K}^b}}\right). \tag{4.22}$$

It is sufficient to use in engineering calculations the gamma-function tables, available, for example, in [25, 46, 52, 53].

The expressions (4.20) and (3.41) indicate that the probabilistically justified safety factor of the element of a serial system in the case of combined loading can be determined by the formula:

$$\overline{K}_\gamma = \left(\frac{n \ln \omega_o t_\gamma}{\ln {}^{1}\!/\!_{\gamma}}\right)^{1/b}, \quad \text{when } \omega_o t_\gamma \geq 10. \tag{4.23}$$

It follows from the expressions (4.15) and (4.23) that to be constant, i.e. such as for one element, the reliability level of a serial system, as the number of its elements increases, it is necessary to increase the safety factors of the elements by $n^{1/b}$ times. For example, if one element $\overline{K} = 1.8$, and the variation coefficients of load and limit load $V = 0.1 (b = 12.15)$, then in order to provide the same reliability function for a serial system of 50 elements, it is necessary to increase the safety factor of the elements by $50^{\frac{1}{12.15}} = 1.38$ times, i.e. to bring them to the level: $\overline{K}_\gamma = 1.8 \cdot 1.38 = 2.48$. An effective way to reduce an effect of the number of elements in a serial system on its reliability is to reduce the level of random scattering of load and limit load. So, if we reduce their variation coefficients to $V = 0.06$, which corresponds (see Table 1.1) $b = 20.68$, then under the conditions of the considered example, the value of the required safety factor for the elements of the system can be reduced to $\overline{K}_\gamma = 1.8 \cdot (50)^{\frac{1}{20.68}} = 2.17$. As at $V \to 0$ a parameter $b \to \infty$, if the elements of the serial system are lack or negligible random scattering of the limit load, the value of the probabilistically justified safety factor actually ceases to be dependent on the number of elements in the system.

Transition to unit distributions. Let us consider a serial system consisting of n elements with the same limit load distribution functions: $G_i(P) = G(P), \quad i = 1,\ 2,\ \ldots,\ n$. Then, in accordance with (4.1), the distribution function of the limit load of the system $G_c(P)$ is determined by the formula:

$$G_c(P) = 1 - (1 - G(P))^n. \tag{4.24}$$

After differentiating with respect to a variable K_c, we obtain.

$$dG_c = n(1 - G)^{n-1} dG. \tag{4.25}$$

In the expression (4.4) for the reliability function of the system for a combined random load, identical for all elements, in accordance with the expression (4.25) and transformation (2.15), we pass to the same-for-all-elements-of-the-system function of the unit distribution of the total load $F_1(G)$ on the system. As a result of this transition, we obtain an expression for the reliability function of a serial system in the form:

$$R_c(m, n) = n \int_0^1 F_1^m(G)(1 - G)^{n-1} dG. \tag{4.26}$$

Another variant of the transition to the unit load distribution function for a system of the form $F_1(G)$ subject to (4.24) can be performed by analogy with the first variant of formulas (2.20) and obtain the equivalent expression (4.26).

$$\begin{aligned} R_c(m, n) &= 1 - m \int_0^1 [1 - (1 - G)^n](F_1(G))^{m-1} f_1(G) dG \\ &= m \int_0^1 (F_1(G))^{m-1} (1 - G)^n f_1(G) dG. \end{aligned} \tag{4.27}$$

If we use the unit distribution function of the limit load of the elements of a system $G_1(F)$ of the form (2.17), then, taking into account (4.24) and (4.25) by analogy with (2.19) and (2.20), we obtain two more versions of equivalent formulas for the reliability function of a serial system:

$$R_c(m,n) = n\int_0^1 F^m[1 - G_1(F)]^{n-1} g_1(F)dF. \tag{4.28}$$

$$R_c(m,n) = m\int_0^1 F^m[1 - G_1(F)]^n F^{m-1}dF. \tag{4.29}$$

The advantage of expressions (4.26 ÷ 29) in predicting the reliability of serial systems from the same type of equally loaded elements is their universality and simplified numerical realizability of computer calculations, which are practically always achievable regardless of the type and parameters of the given load distributions and limit load of the elements. The numerical results obtained with the four formulas given above must coincide, and the choice between them can be determined by convenience considerations in obtaining analytical dependencies.

For example, in the case of general load distributions and bearing capacities of elements, according to the Frechet law (2.54) with the same shape parameters $\rho_{\mathrm{H}} = \rho_{\Pi} = \rho$ in accordance with (2.59) from (4.26), we obtain an expression for the reliability function of a serial system:

$$R_c(m,n) = n\int_0^1 G_{m/\overline{K}^\rho}(1 - G)^{n-1}dG. \tag{4.30}$$

The integral in (4.30) is defined as the beta function [46] and similarly to the expression (2.31) for the reliability function of the system, we obtain the formula

$$R_c(m,n) = n \cdot B\left(n,\ 1 + \frac{m}{\overline{K}^\rho}\right). \tag{4.31}$$

Taking into account that n is always an integer and expressing, in analogy with (2.32), a beta function by means of gamma functions, we finally express (4.31) in the form

$$R_c(m,n) = \prod_{i=1}^{n} \frac{i\overline{K}^\rho}{i\overline{K}^\rho + m}. \tag{4.32}$$

The comparison of formulas (4.32) and (4.12), when, it is necessary to put $\chi = n$ and $\overline{K}_{\min} = \overline{K}$ if the loaded elements are equal, shows that the arguments m and n in the expressions for the reliability function of the serial system change places if the Weibull distribution of load and limit load is replaced by the Fréchet distribution.

Proceeding from (4.32), we can use the approximate formula by analogy with (4.14) at $n > 5$ to determine the probabilistically justified safety factor:

$$\overline{K}(m,\gamma) = \left\{ \frac{m[C_o + \ln(n + 0.51)]}{\ln 1/\gamma} \right\}^{1/\rho}, \quad (4.33)$$

where $C_o = 0.57721\ldots$

In order to estimate the probabilistically justified safety factors for the elements of serial systems the presence of two expressions (4.15) and (4.33) makes it possible to perform calculations in each case, starting from known values of n, m, b, ρ and γ, and choose a larger value $\overline{K}(m, \ \gamma)$, taking into account the uncertainty regarding a possible type of load distributions and limit load.

In the case of a Poisson stream of extreme loads acting jointly on a serial system that consists of n elements with the same limit load functions, using (4.26), one can obtain an expression for the conditional probability of the system's non-failure operation for any random number i of extreme loads:

$$R_c(i,n) = n \int_0^1 F_1^i(G)(1-G)^{n-1}dG. \quad (4.34)$$

Taking into account (1.37) and (1.38), the expression for the reliability function of the system as a function of the operating time for a stationary Poisson flow of combined loadings has the form:

$$R_c(t,n) = e^{-\omega_o t} \sum_{i=0}^{\infty} \frac{(\omega_o t)^i}{i!} R_c(i,n). \quad (4.35)$$

Substituting (4.34) into (4.35), after transformations analogous to those that were performed for the derivation of formulas (3.31), we obtain a general expression for the reliability function of a serial system, depending on the operating time:

$$R_c(t,n) = n \int_0^1 e^{-\omega_o t[1-F_1(G)]}(1-G)^{n-1}dG. \quad (4.36)$$

Forecasting the reliability of systems with the help of the expression (4.36) it is convenient to implement on the computer, setting different functions of the unit load distribution $F_1(G)$. A set of such functions, corresponding to different combinations of load distributions and limit load, is given in the appendixes (see Table A.2). Some analytical results can also be obtained with the help of (4.36). Thus, it is easy to verify that by specifying $F_1(G) = 1 - (1-G)^{\overline{K}^b}$, from (4.36), by changing the variable $x = (1-G)^{\overline{K}^b}$, we can obtain the formula (4.21).

On the basis of using the unit distribution function of the limit load of elements$G_1(F)$, taking into account that the conditional reliability function with a random number i of loads is expressed by the formula:

$$R_c(i,n) = n\int_0^1 F^i[1 - G_1(F)]^{n-1}g_1(F)dF, \tag{4.37}$$

under the Poisson flow of extreme loads, one can obtain a dependence equivalent to the expression (4.36) in the following form

$$R_c(t,n) = n\int_0^1 e^{-\omega_o t(1-F)}[1 - G_1(F)]^{n-1}g_1(F)dF. \tag{4.38}$$

With a common (combined) method of random loading of elements in the serial system, an alternative is possible when the limit load of all elements has negligible dispersion, i.e. The limit load of each element is practically constant. In this case, sudden failures of the system under extreme loads will always be due to failures of one or several elements having the lowest deterministic limit load, which can be called limiting. Then the reliability function of the system will not depend on the total number of elements included in it, but will be determined only by the reliability function of the limiting elements $R_{\min}(t)$. The probability $R_{\min}(t)$ can be estimated by the methods described in Chap. 1, and the reliability function of a serial system containing limiting elements $R_c(t) \approx R_{\min}(t)$.

Separately independent loading of elements in the system. Taking the scheme of separately independent loading of elements when predicting the system reliability indicators, it is necessary to consider that failures or operable states of the elements are independent random events among themselves. Then the reliability function of a serial system $R_c(t)$ should be determined as the product of faultness probabilities of its elements:

$$R_c(t) = \prod_{i=1}^{n} R_i(t). \tag{4.39}$$

Using the basic relationship [47] between the reliability function and the failures rate (risk): $R_i(t) = e^{-\int_0^t \lambda_i(t)dt}$, based on (4.39) we can obtain an expression relating the failure rate of a serial system to the failure rates $\lambda_i(t)$ of the elements entering into it:

$$\lambda_c(t) = \sum_{i=1}^{n} \lambda_i(t). \tag{4.40}$$

In the case of constant deterministic values of the limit load of the system elements, their failure rates in accordance with (1.41) for stationary Poisson flows of extreme loads will not depend on the operating time, and the failure rate of the ith element can be determined by the formula:

$$\lambda_i = (1 - R_1(K_i))\omega_{oi}, \quad i = 1, 2, \ldots, n, \tag{4.41}$$

where $K_i = \frac{P_{oi}}{\overline{P}_{Hi}}$—a safety factor; P_{oi}—a limiting deterministic level of the limit load of the ith element; $\overline{P}_{Hi}$—an average value of the extreme load acting on the element; ω_{oi}—an intensity of element loads; $R_1(K_i)$—a reliability function of the ith element at the first loading, determined by the formulas of Table 1.2 depending on the type of load distribution.

Then the constant value of the failure rate of a serial system is defined as the sum (4.40) of the failure rates of its elements:

$$\lambda_c = \sum_{i=1}^{n} (1 - R_1(K_i))\omega_{oi}, \tag{4.42}$$

and the reliability function of the system γ

$$R_c(t) = e^{-\lambda_c t}, \tag{4.43}$$

i.e. the operating time to failure of a serial system under the Poisson's separately independent loading of elements with deterministic limit load is subject to exponential distribution.

The mean time to failure of such a system is determined from the expression:

$$T_c = \left(\sum_{i=1}^{n} \frac{1 - R_1(K_i)}{T_{oi}} \right)^{-1}, \tag{4.44}$$

where T_{oi} is an average period between loads of the ith element.

In the case where the system consists of the same equally loaded elements, when $T_{oi} = T_o$ and $K_i = K$; $i = 1, 2, \ldots, n$, then for separately independent loading of elements the mean time to failure is calculated by the formula:

$$T_c = \frac{T_o}{n(1 - R_1(K))} = \frac{T}{n}, \tag{4.45}$$

where T—the mean time to failure of one element.

Based on (1.42) and (4.45), one can use the data of the application Table A.1 to predict the gamma-percentile operating time to failure $t_\gamma(K)$ of a serial system of n identical elements happens. In this case, the tabulated values $\tau_\gamma(K)$ should be multiplied by the ratio T_o/n.

If the limit load of the system elements has random scattering, and the loading of the elements in the system is assumed to be separately independent, then the expressions (4.39) and (4.40) for the reliability function and the failure rate of the serial system remain valid. Then, in accordance with (3.31) and (4.39), the reliability function of the system in the case of Poisson load flows can be predicted with the expression:

$$R_c(t) = e^{-t\sum_{i=1}^{n} \omega_{oi}} \prod_{i=1}^{n} \int_0^1 e^{t\omega_{oi}F_{1i}(G)} dG, \tag{4.46}$$

where ω_{oi}—an intensity of loads of elements;
$F_{1i}(G)$—functions of unit load distributions.

The function of the system failure rate in accordance with (3.33) and (4.40) can be determined by the formula:

$$\lambda_c(t) = \sum_{i=1}^{n} \omega_{oi} \left(1 - \frac{\int_0^1 F_{1i}(G) e^{t\omega_{oi}F_{1i}(G)} dG}{\int_0^1 e^{t\omega_{oi}F_{1i}(G)} dG} \right). \tag{4.47}$$

In the case of random scattering of the limit load of the elements, the summands of the sum (4.47) are monotonically decreasing operating time functions. Therefore, the failure rate of a serial system will also decrease monotonically, and the time between sudden failures in the system will not obey an exponential law of the form (4.43). In such systems, it may be effective to conduct a preliminary break-in with the "burning out" of the weakest elements and an increase in the reliability level.

If a serial system consists of identical equally loaded elements, then for separately independent loads, the expressions for the reliability function and the rate of sudden failures take the form:

$$R_c(t, n) = e^{-nt\omega_o} \left(\int_0^1 e^{t\omega_o F_1(G)} dG \right)^n. \tag{4.48}$$

$$\lambda_c(t) = n\omega_o \left(1 - \frac{\int_0^1 F_1(t) e^{t\omega_o F_1(G)} dG}{\int_0^1 e^{t\omega_o F_1(G)} dG} \right). \tag{4.49}$$

In order to determine the reliability function of the elements, one can use the analytic relations (3.38) and (3.39). Then, the prediction of the reliability function of a serial system with separately independent loads of elements is performed using the expression:

$$R_c(t,n) = \left(\frac{\Gamma\left(1 + {}^{1}\!/\!_{K^b}\right) - \Gamma\left(1 + {}^{1}\!/\!_{K^b}, \omega_o t\right)}{(\omega_o t)^{1/K^b}} + e^{-\omega_o t} \right)^n. \tag{4.50}$$

A more convenient approximate dependence, which can be recommended to use at $\omega_o t \geq 10$, has the form:

$$R_c(t,n) = \frac{\Gamma^n\left(1 + {}^{1}\!/\!_{K^b}\right)}{(\omega_o t)^{n/K^b}}. \tag{4.51}$$

The use of formulas (4.50) and (4.51) assumes that the random bearing capacities and loads acting on the elements have the same variation coefficient and are distributed according to the Weibull law. Otherwise, to predict the reliability function of the system, one can use the general expression (4.39), where each multiplier R_i should be determined from the expressions given in Chaps. 2 and 3 and obtained with various combinations of the laws of the load distribution and limit load of the elements.

Comparison and analysis of the results obtained by formulas (4.21) and (4.22) with the simultaneous loading of the elements of a serial system with those that can be obtained by separately independent loading with (4.50) and (4.51) allows us to draw the following general conclusion:

Other things being equal, the separately independent extreme loads of the elements of a serial system reduce its reliability in comparison with the combined loading.

It is not always possible to ensure the combined loading of all elements in the system. Therefore, in order to ensure the reliable prediction of the reliability function of serial systems, it is advisable to use a scheme of separately independent loading of elements, in accordance with which the expressions (4.39) and (4.46) are the main ones in predicting.

4.2 Fault-Tolerant System

The fault-tolerance property of a system is related to its structure and a concept of a system failure. It is necessary to classify systems whose failure occurs as a result of simultaneous failure of more than one element to the category of fault-tolerant or having a reserve of survivability systems. The concept of an "uncritical" failure of an element in a fault-tolerant system is sometimes interpreted as a refusal restored within an acceptable period of time or associated with a certain but limited material damage. A sudden mechanical failure may happen in a fault-tolerant system due to the loss of the limit load as a result of a simultaneous sudden failure of more than one of its elements. One will not take into account the issues of recoverability of failures or other negative consequences. A value essential and determining a system failure will be only a number of simultaneously failed due to the loss of the limit load of elements in the future consideration. Proceeding from this, it is necessary to analyze a possibility of practical application of such a structural scheme in each specific case.

A particular case of fault-tolerant systems is systems with a parallel structure, which fail only when all elements fail. If the elements of a parallel system are loaded separately and independently, then failures of such a system can be considered as independent random events, and the reliability function can be determined with the expression:

$$\widetilde{R}_c(t) = 1 - \prod_{i=1}^{n} (1 - R_i(t)), \tag{4.52}$$

where $R_i(t)$—reliability function functions depending on the operating time.

Consider the case where the limit load of the elements in a parallel system does not have random scattering and are deterministic constant values. Then for stationary Poisson flows of elements with intensities ω_{oi} in accordance with the results of Chap. 1, based on (1.39) and (4.52), we obtain the following expression for the reliability function of the system:

$$\widetilde{R}_c(t) = 1 - \prod_{i=1}^{n} \left(1 - \exp\left[-\omega_{oi} t(1 - R_{1i}(K_j))\right]\right). \tag{4.53}$$

If the intensity of loads and the safety factor are the same for all elements of the system, i.e. $\omega_{oi} = \omega_o$, $K_i = K$ $i = 1, 2, \ldots, n$; then we obtain from (4.53) that

$$\widetilde{R}_c(t, n) = 1 - (1 - \exp\left[-\omega_o t(1 - R_1(K))\right])^n. \tag{4.54}$$

The mean time to failure of the system with a parallel structure consisting of equidistant elements, can be determined from the expression:

$$T_c = \int_o^\infty \widetilde{R}_c(t)dt = \int_o^\infty \{1 - (1 - \exp\left[-\omega_o t(1 - R_1(K))\right])^n\}dt. \tag{4.55}$$

The expression (4.55) can be integrated in a closed form [47] by changing the variable

$$x = 1 - \exp\left[-\omega_o t(1 - R_1(K))\right]. \tag{4.56}$$

It leads this integral to the form:

$$T_c = \frac{1}{\omega_o(1 - R_1(K))} \cdot \int_o^1 \frac{1 - x^n}{1 - x} dx. \quad (4.57)$$

After integrating, we finally obtain that

$$T_c = \frac{\sum_{i=1}^{n} \frac{1}{i}}{\omega_o(1 - R_1(K))}. \quad (4.58)$$

The expression (4.58) indicates the possibility and shows a way to management the reliability of the parallel system due to the size of the safety factor of the elements. To do this, use the formulas given in Table 1.2 depending on a type of load distribution. Subject to (1.40) and (4.58), the dependence between the mean time to failure of a system with a parallel structure and the mean time to failure T of its elements for separately independent loading takes the following form:

$$T_c = T\left(1 + \frac{1}{2} + \ldots + \frac{1}{n}\right). \quad (4.59)$$

The comparison of the expressions (4.59) and (4.45) shows the difference in the influence of the number of elements on the reliability index of the system T_c, depending on its structure.

If the limit load of elements has random scattering, and the loading method is separately independent, the expression (4.52) to predict the reliability function of a system with a parallel structure remains valid. In this case, the reliability function of the elements $R_i(t)$ is determined using the methods given in Chap. 3. The general expressions for $R_i(t)$ are the formulas (3.30) and (3.31). On their basis the analytical dependences (3.38), (3.39), and (3.43) are obtained.

For example, using (4.52) and (3.38), one can obtain an analytical expression to predict the reliability function of a system with a parallel structure in the case when extreme loads and the bearing capacities of the elements have the Weibull distribution:

$$\widetilde{R}_c(t) = 1 - \prod_{i=1}^{n} \left\{ 1 - e^{-\omega_{oi} t} - \frac{\Gamma\left(1 + 1/\overline{K}_i^b\right) - \Gamma\left(1 + 1 + 1/\overline{K}_i^b,\ \omega_{oi} t\right)}{(\omega_{oi} t)^{1/\overline{K}_i^b}} \right\}. \quad (4.60)$$

In the case of equally loaded elements, when $\omega_{oi} = \omega_o$ and $\overline{K}_i = \overline{K}\, i = 1, 2, \ldots, n$; when $\omega_o t \geq 10$ we have an approximate expression.

$$\widetilde{R}_c(t,n) = 1 - \left(1 - \frac{\Gamma\left(1 + {}^{1}\!/\!_{K^b}\right)}{(\omega_o t)^{1/K^b}}\right)^n. \tag{4.61}$$

Let's consider a case of combined extreme loading and an appropriate scheme of a sudden mechanical failure of a system with a parallel structure. A sudden failure of a system with a parallel structure under simultaneous quasistatic loading will occur when the extreme load in all elements at the same time exceeds their limit load. It is assumed that before the failure of the last element in the system, the short-time limit load of the remaining elements is not accompanied by a discontinuous redistribution of the load between the elements. Such a situation is possible, for example, under the conditions of the elastoplastic state of the material, if the resistance to deformation for all simultaneously loaded elements of the system does not decrease until the load exceeds the limit load of the last element.

If the limit load of the elements of a parallel system is a deterministic value, then under simultaneous extreme loading of the elements, sudden system failures will occur as a result of the simultaneous excess of the limit load of those elements that have the highest deterministic level of limit load. In this case, if we exclude the possibility of an abrupt load transfer between elements in the loading process, which can be caused by failures of more "weak" elements, we can assume that the reliability function of a parallel system will be determined by the reliability function of its most reliable elements, i.e. $\widetilde{R}_c(t) \approx R_{\max}(t)$.

Next, let us consider how to build a system probabilistic reliability model a with a parallel structure (Fig. 4.1b) for repeated and identical for all elements random combined extreme loading and random limit load of elements. If functions of the limit load distributions are given: $G_i(P_{\Pi i})$, $i = 1, 2, \ldots, n$;, and the distribution function of the total load on the elements $F(P)$, then the limit load distribution function of the parallel system is given by

$$\overline{G}_c(P_\Pi) = \prod_{i=1}^{n} G_i(P_{\Pi i}), \tag{4.62}$$

and the distribution function of the load maximum for m-fold loading:

$$F(P_{\max}) = F^m(P). \tag{4.63}$$

While differentiating (4.62), we obtain.

$$d\widetilde{G}_c(P_\Pi) = \widetilde{g}_c(P_\Pi)dP_\Pi = d\left(\prod_{i=1}^{n} G_i(P_{\Pi i})\right), \tag{4.64}$$

where $\widetilde{g}_c(P_\Pi)$—a density distribution of the limit load of a parallel system.

The differentiation (4.63) gives

$$dF(P_{\max}) = mF^{m-1}(P)f(P)dP, \tag{4.65}$$

where $f(P)$—a density of the total load distribution;

Then the reliability function of a parallel system under simultaneous m-fold extreme loading of its elements by analogy with (4.3) and (4.4) can be determined from the expression:

$$\widetilde{R}_c(m) = \int_o^\infty F^m(P)\widetilde{g}_c(P)dP = \int_o^\infty F^m(P)d\left(\prod_{i=1}^n G_j(P)\right). \tag{4.66}$$

Using (4.62) and (4.65), we can also determine the failure probability of a parallel system under combined loading

$$\widetilde{Q}_c(m) = m\int_o^\infty \left(\prod_{i=1}^n G_j(P)\right)F^{m-1}(P)f(P)dP. \tag{4.67}$$

Therefore, it is possible, when convenient, to apply one more version of the probability model for the parallel system's reliability function:

$$\widetilde{R}_c(m) = 1 - \widetilde{Q}_c(m) = 1 - m\int_o^\infty \left(\prod_{i=1}^n G_i(P)\right)F^{m-1}(P)f(P)dP. \tag{4.68}$$

It is convenient to analyze the reliability of a system with a parallel structure under repeated combined loading in the case when the load and the limit load have the Frechet distribution. Then, if the distribution function of the limit load of any ith element has the form

$$G_i(P) = \exp\left[-\left(\frac{h_{\Pi i}}{P}\right)^\rho\right],$$

And the distribution function of the total load is given by

$$F(P) = \exp\left[-\left(\frac{h_{\text{H}}}{P}\right)^\rho\right],$$

Then, based on (4.67), the fault probability of a parallel system consisting of n elements for their simultaneous m-fold loading can be determined as follows:

$$\widetilde{Q}_c(m) = \int_0^\infty \frac{m\rho}{P}\left(\frac{h_H}{P}\right)^\rho \exp\left[-\left[\frac{\sum_{i=1}^{n} h_{\Pi i}^\rho + mh_H^\rho}{P^\rho}\right]\right] dP. \quad (4.69)$$

If (4.69) is first multiplied and divided by the ratio

$$\frac{\sum_{i=1}^{n} h_{\Pi i}^\rho + mh_H^\rho}{mh_H^\rho}.$$

Then, it is integrated. Taking into account (4.68), the result is an analytical expression for the reliability function of the parallel system:

$$\widetilde{R}_c(m) = \frac{\sum_{i=1}^{n} \overline{K}_i^\rho}{\sum_{i=1}^{n} \overline{K}_i^\rho + m}, \quad (4.70)$$

where $\overline{K}_i = \frac{h_{\Pi i}}{h_H}$ is the safety factor of the ith element.

Transforming this expression, we have the following formula:

$$\overline{R}_c(m) = \frac{\overline{K}^\rho_{max}\chi_\Pi}{\overline{K}^\rho_{max}\chi_\Pi + m}, \quad (4.71)$$

where $\overline{K}_{max}$—a safety factor for the least loaded element in the system;

$\chi_\Pi = \sum_{i=1}^{n}\left(\frac{\overline{K}_i}{\overline{K}_{max}}\right)^\rho$—a coefficient that ensures the reduction of the elements' number in the system to the least loaded.

The value χ_Π is within the interval $1 < \chi_\Pi \leq n$ and in the case of equally loaded elements in a parallel system, when all $\overline{K}_i = \overline{K}$ we have $\chi_\Pi = n$. Then it follows from (4.71) that

$$\widetilde{R}_c(m, n) = \frac{n\overline{K}^\rho}{nK^\rho + m}. \quad (4.72)$$

Based on (4.71), it is easy to obtain a formula to calculate the safety factor of the least loaded element, which provides a predetermined value γ of the reliability function of a parallel system under repeated loading.

$$\overline{K}_{max}(\gamma) = \left[\frac{m\gamma}{\chi_\Pi(1-\gamma)}\right]^{1/\rho}. \quad (4.73)$$

Using the mean value of the safety factor of the elements in the system $\overline{\overline{K}} = \sum^{n} \overline{K}_i/_n$, the expression (4.71) can be rewritten in the form.

$$\widetilde{R}_c(m) = \frac{\overline{\overline{K}}^{\rho} \overline{\chi}_{\Pi}}{\overline{\overline{K}}^{\rho} \overline{\chi}_{\Pi} + m}, \tag{4.74}$$

where $\overline{\chi}_{\Pi} = \sum_{i=1}^{n} \left(\frac{\overline{K}_i}{\overline{\overline{K}}} \right)^{\rho}$—a coefficient that ensures the reduction of the elements' number in the system to the conditional mean.

Taking into account $\overline{\chi}_{\Pi} \geq n$, it follows from (4.74) that the minimum value of the reliability function of a parallel system when the condition $\overline{\overline{K}} = \text{const}$ is satisfied is.

$$\min \widetilde{R}_c(m) = \frac{\overline{\overline{K}}^{\rho} n}{\overline{\overline{K}}^{\rho} n + m}.$$

This is the case only when all the elements of the system are equally loaded, i.e. all $\overline{K}_i = \overline{\overline{K}}$. Consequently, unlike serial systems, in the case of combined loading of elements of a system with a parallel structure, equal loading is undesirable, since keeping the mean value of the safety factors results in minimizing the system's reliability function. The most effective way to increase the level of reliability of such systems is to increase the safety factor for the least loaded elements.

It makes sense to numerically compare the reliability prediction of a parallel system with the separately independent method of loading elements using the formula (4.52) with the results of predicting the reliability of a system from the same elements, but with their combined loading. This alternative can be analyzed using the expression (4.72). Let us consider a parallel system of two identical elements, where the value $\overline{K}^{\rho} = 200$ that with the variation coefficient of the load and limit load $V = 0.1$ corresponds to the shape parameter $\rho = 13.62$ and the safety factor $\overline{K} = 1.4755$. If the given number of loads is $m = 10$, then the reliability function of one element is determined by the formula (2.60), hence it follows that

$$R_{10} = \frac{200}{200 + 10} = 0,95238.$$

Then the predicted reliability function of a parallel system for separately independent loading of elements, calculated by the formula (4.52), is:

$$\widetilde{R}_c = 1 - (1 - 0.95238)^2 = 0.99773.$$

This calculation shows that the effect of the supposed "duplication" of elements in the system with separately independent loading can be significant. However, if we consider the combined loading of the elements, then using (4.72) we obtain

$$\widetilde{R}_c(10) = \frac{2 \cdot 200}{2 \cdot 200 + 10} = 0.97561$$

and the difference in the reliability predictions of the same parallel system, depending on the method of loading of its elements, is significant.

Let us consider the case of the combined repeated random loading of a system with a parallel structure for all elements (Fig. 4.1b) when the limit load of all its elements have the same distributions with the function *G*(*P*). Then it follows from (4.66) that the reliability function of the system will be determined by the expression

$$\widetilde{R}_c(m,n) = n\int_0^{\infty} F^m(P)G^{n-1}(P)g(P)dP, \tag{4.75}$$

where *g*(*P*) is a density of distribution of the limit load of the elements.

Having transited in (4.75) to unit distributions, we obtain that

$$\widetilde{R}_c(m,n) = n\int_0^1 F_1^m(G)G^{n-1}dG. \tag{4.76}$$

Another case of the model to predict the reliability function of a parallel system aftertransition to unit distributions can be obtained from (4.68) in the form.

$$\widetilde{R}_c(m,n) = 1 - m\int_0^1 G_1^n(F)F^{m-1}dF. \tag{4.77}$$

If the load and limit load of elements are distributed according to the Weibull law and are similar (have the same parameter b), the function of the unit load distribution is given by: $F_1(G) = 1 - (1 - G)^{\overline{K}^b}$. Then it follows from (4.76) that

$$\widetilde{R}_c(m,n) = n\int_0^1 \left[1 - (1 - G)^{\overline{K}^b}\right]^m G^{n-1}dG. \tag{4.78}$$

The substitution of the variable $x = 1 - (1 - G)^{\overline{K}^b}$ leads the integral (4.78) to the form.

$$\widetilde{R}_c(m,n) = \frac{n}{\overline{K}^b}\int_0^1 x^m\left[1 - (1 - x)^{1/\overline{K}^b}\right]^{n-1}(1 - x)^{1/\overline{K}^b - 1}dx. \tag{4.79}$$

Using the binomial expansion

$$\left[1-(1-x)^{1/\overline{K}^b}\right]^{n-1}=\sum_{i=0}^{n-1}(-1)^i\binom{n-1}{i}(1-x)^{i/\overline{K}^b},$$

where $\binom{n-1}{i}=\frac{(n-1)!}{i!(n-1-i)!}$—binomial coefficients.

From (4.79) we find that

$$\widetilde{R}_c(m,n)=\frac{n}{\overline{K}^b}\sum_{i=0}^{n-1}(-1)^i\binom{n-1}{i}\int\limits_0^1 x^m(1-x)^{\frac{i+1}{\overline{K}^b}-1}dx. \tag{4.80}$$

Integrals in the sum (4.80) are beta functions, therefore

$$\widetilde{R}_c(m,\ n)=\frac{n}{\overline{K}^b}\sum_{i=0}^{n-1}(-1)^i\binom{n-1}{i}\frac{\Gamma(m+1)\Gamma\left(\frac{i+1}{\overline{K}^b}\right)}{\Gamma\left(m+1+\frac{i+1}{\overline{K}^b}\right)}. \tag{4.81}$$

Having performed the transformations in (4.81) analogous to when deriving the formula (2.34), we obtain an expression for the reliability function of a parallel system for an integer number m of combined extreme loads of its elements:

$$\widetilde{R}_c(m,n)=\sum_{i=0}^{n-1}(-1)^i\binom{n}{i+1}\prod_{j=1}^{m}\frac{j\overline{K}^b}{j\overline{K}^b+i+1}, \tag{4.82}$$

where $\binom{n}{i+1}=\frac{n!}{(i+1)!(n-1-i)!}$.

In order to apply the formula (4.82) for large m in engineering calculations, it is convenient to use the expression (2.39) and obtain from (4.82) an approximate formula:

$$\widetilde{R}_c(m,n)=\sum_{i=0}^{n-1}(-1)^i\binom{n}{i+1}\exp\left\{-\Omega_i S_1(m)+\frac{\Omega_i^2}{2}S_2(m)-\frac{\Omega_i^3}{3}S_3(m)\right\}, \tag{4.83}$$

where $\Omega_i=\frac{i+1}{\overline{K}^b}$; $i=0,1,\ldots,n-1$. The values of the coefficients $S_1(m)$, $S_2(m)$ and $S_3(m)$ are given in Table 2.1.

In the case when $m>100$ the following approximate expression can be used on the basis of (2.41):

$$\widetilde{R}_c(m,n) = \sum_{i=0}^{n-1} (-1)^i \binom{n}{i+1} (m+0.51)^{-\Omega_i} \exp\left(-0.57721\Omega_j + \frac{\pi^2}{12}\Omega_i^2 - 0.4\Omega_i^3\right). \tag{4.84}$$

Comparing expressions (4.76) and (4.26) and taking into account that

$$(1-G)^{n-1} = \sum_{i=0}^{n-1} (-1)^i \binom{n-1}{i} G^i.$$

It can be shown that with the same combination of combined load distributions and limit load of the elements, the general dependence between the reliability function of a parallel system and the reliability function of the corresponding (composed of the same elements) serial systems is valid:

$$\widetilde{R}_c(m,\ n) = \sum_{i=0}^{n-1} (-1)^i \binom{n}{i+1} R_c(m,\ i+1). \tag{4.85}$$

It is obvious that the formula (4.82) is a particular case of the expression (4.85) for the load distribution and the limit load of elements according to the Weibull law.

Figure 4.3 shows the graphs of the dependences of the reliability function on the number of combined extreme loads, obtained (solid lines) with the help of formulas (4.82) and (4.83) for the element (for $n = 1$) and parallel systems ($n = 2$, 3) for $\overline{K} = 1.2$ and $b = 10$. Here, the graphs of the reliability function of parallel systems for separately independent extreme loading of elements are shown with the dashed lines and obtained with the help of expression (4.52).

The graphs in Fig. 4.3 show that under repeated extreme loading of parallel systems in the sense of reliability, combined loading of elements leads to a decrease in the reliability function in comparison with the case of the separately independent loading. To ensure independent extreme loading of each element in the system in practice is problematic, it is expedient to use the expressions (4.71) and (4.83) for a guaranteed prediction of the reliability of parallel systems.

From the above, the conclusion follows that, other things being equal, the combined extreme loading of elements in a parallel system leads to a decrease in its reliability in comparison with the case of separately independent loading.

It should also be noted that the use of the Frechet distribution to predict of the reliability of parallel systems and, accordingly, the application of formulas (4.71) and (4.72), gives greater guarantees of overestimation of reliability than the use of formulas (4.82) and (4.83), which correspond to the Weibull distribution.

Considering the case of the combined extreme loading of a system with a parallel structure under a Poisson time load of the form (1.37) and using (4.76), we obtain an expression analogous to the expression (4.34) for the conditional reliability function of the system if it is acted upon by a random number i of loads:

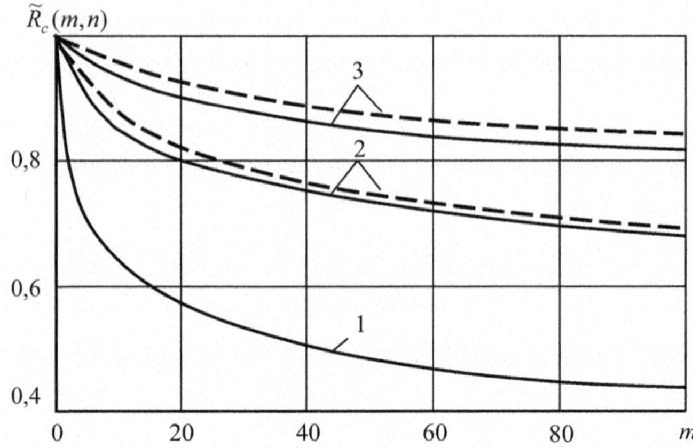

Fig. 4.3 Graphs of the dependencies of reliability function on the number of extreme loads with: (*1*) $n = 1$; (*2*) $n = 2$; (*3*) $n = 3$.

$$\widetilde{R}_c(i,n) = n\int_0^1 F_1^i(G)G^{n-1}dG. \tag{4.86}$$

Then, based on (1.38), the reliability function of a parallel system for a steady Poisson load flow, depending on the operating time and the number of equally loaded elements, is given by the following:

$$\widetilde{R}_c(t,n) = e^{-\omega_o t} n \sum_{i=0}^{\infty} \frac{(\omega_o t)}{i!} \int_0^1 F_1^i(G)G^{n-1}dG, \tag{4.87}$$

which after the transformations takes the form:

$$\widetilde{R}_c(t,n) = n\int_0^1 e^{\omega_o t[1-F_1(G)]}G^{n-1}dG. \tag{4.88}$$

The comparison of the formulas (4.88) and (4.36) makes it possible to obtain a similar dependence (4.85) connecting the reliability function of parallel and serial systems consisting of the same elements with the same combination of the load distributions and limit load:

$$\widetilde{R}_c(t,n) = \sum_{i=0}^{n-1} (-1)^i \binom{n}{i+1} R_c(t,\ i+1). \tag{4.89}$$

In the particular case of the distribution of the total load on the system and the limit load of the equally loaded elements according to the Weibull law with the same parameter of form b, it follows from (4.21) and (4.89) that for a steady Poisson load flow the reliability function of a parallel system can be determined from the expressions:

$$\widetilde{R}_c(t,n) = \sum_{i=0}^{n-1} (-1)^i \binom{n}{i+1} \left\{ \frac{\Gamma\left(1 + \frac{i+1}{\overline{K}^b}\right) - \Gamma\left(1 + \frac{i+1}{\overline{K}^b}, \omega_o t\right)}{(\omega_o t)^{(i+1)/\overline{K}^b}} \right\} + e^{\omega_o t}, \quad (4.90)$$

where the formula (3.39) can be used to calculate incomplete gamma functions $\Gamma\left(1 + \frac{i+1}{\overline{K}^b}, \omega_o t\right)$.

If the dimensionless operating time is $\omega_o t > 10$, then the reliability function of the parallel system can be predicted using an approximate expression:

$$\widetilde{R}_c(t,n) = \sum_{i=0}^{n-1} (-1)^i \binom{n}{i+1} \frac{\Gamma\left(1 + \frac{i+1}{\overline{K}^b}\right)}{(\omega_o t)^{(i+1)/\overline{K}^b}}. \quad (4.91)$$

In addition to (4.88) for the case of a stationary Poisson flow of the combined loads of a parallel system using the function of a unit distribution of the limit load $G_1(F)$, one more expression can be obtained that determines the system's reliability function:

$$\widetilde{R}_c(t,n) = n \int_0^1 e^{-\omega_o t(1-F)} G_1^{n-1}(F) g_1(F) dF. \quad (4.92)$$

Expressions (4.88) and (4.92) are equivalent and while doing computer calculations, for example, using Mathcad, they can equally be used for various combinations of load distributions and limit load of elements. In this case, it is expedient to apply expressions for the functions of the unit distributions $F_1(G)$ and $G_1(F)$, which are given in the appendices (see Table A.2) and directly depend on the safety factors $\overline{K}$. This simplifies the reliability management of systems due to a reasonable choice of safety factors for elements.

Turning to the consideration of a more general case of fault-tolerant systems with a loaded survivability reserve (Fig. 4.1c), let us dwell in more detail on the case of separately independent loading of elements in the system. Let us assume that with this method of loading a system consisting of n constructively identical elements after the same number of extreme loads, all elements will be equally reliable, and the reliability function for each is the same and equal $R_э(m)$. To do this, it is sufficient that all the elements have the same load distribution functions $G(P)$ and safety factors $\overline{K}$, determined by the mean load and limit load. In addition, for random sequences of discrete loads (Fig. 1.1b), separately and independently acting on each element, there must be the same distribution function $F(P)$. If the above conditions are met, failures and no-failure operation of elements in the system can be considered independent random events; a binomial distribution can be used for analysis and prediction of its reliability use [56].

If a system consisting of n equally reliable elements has a critical number of failures of the elements leading the system to failure l, and the maximum permissible (ensuring the system's up state) the number of failing elements is $l - 1$, then for the same number of m separately independent loads of all elements in accordance with a binomial distribution, the reliability function of the system can be determined from the expression:

$$\overline{R}_c(l, m, n) = \sum_{i=0}^{l-1} \binom{n}{i} (1 - R_{\text{э}} \tag{4.93}$$

where $\binom{n}{i} = \frac{n!}{i!(n-i)!}$—binomial coefficients.

The expression (4.93) implies [56] that the condition $0 < R_{\text{э}}(m) < 1$ is satisfied. The degree of relative fault tolerance of such a system can be estimated by the survivability reserve $S = \frac{l-1}{n}$.

It is easy to verify that the expression (4.93) is a generalization of formula (4.39) for serial structures (when $l = 1$ and $R_i = R_{\text{э}}$) and the expression (4.52) is for parallel structures (when $l = n$ and $R_i = R_{\text{э}}$) of systems in the sense of reliability.

If the stationary Poisson flows of extreme loads are the same in intensity ω_o and load distribution functions $F(P)$ for each element of the system, then, in accordance with (3.31), the reliability function of the element can be determined from the expression depending on the operating time:

$$R_{\text{э}} \tag{4.94}$$

where $F_1(G)$—a function of the unit load distribution is the same for all elements of the system.

So, for example, in the case of load distribution and limit load of elements according to the Weibull law, the value $R_{\text{э}}(t)$ can be determined by the formula (3.38).

Consequently, in the case of separately independent loading of the elements, the reliability function of a system with a loaded survivability reserve in accordance with (4.93) can be predicted with the help of the expression:

$$\overline{R}_c(l, t, n) = \sum_{i=0}^{l-1} \binom{n}{i} (1 - R_{\text{э}} \tag{4.95}$$

in which the operating time dependent function $R_{\text{э}}(t)$ is determined by the formula (4.94).

In computer calculations with the help of (4.95), for all values of t, it is necessary to ensure strict fulfillment of the condition: $0 < R_3(t) < 1$.

On the basis of (4.93), in view of the independence of random variables of the bearing capacities of the elements among themselves, we can obtain the expression for the limit load distribution function of the system under consideration:

$$G_c(P) = 1 - \sum_{i=0}^{l-1} \binom{n}{i} G^i(P)(1 - G(P))^{n-i}, \tag{4.96}$$

which connects the distribution function of the system's limit load with the distribution function of the limit load of one element $G(P)$.

In the particular case for the consistent system when $l = 1$, the formula (4.24) is obtained from (4.96).

Differentiating expression (4.96) with respect to the variable G, we obtain:

$$dG_c = \sum_{i=0}^{l-1} \binom{n}{i} (nG - i)G^{i-1}(1 - G)^{n-i-1} dG. \tag{4.97}$$

When $l = 1$ (for a consistent system) this expression coincides with (4.25).

Let us now consider a case of the simultaneous and identical m-fold random load of all elements of a system with a loaded survivability reserve, when the function of the unit distribution of the total load $F_1(G)$ is the same for all elements of the system. In this case, based on (4.97), we obtain an expression for the reliability function of the system:

$$R_c(l, m, n) = \sum_{i=0}^{l-1} \binom{n}{i} \int_0^1 F_1^m(G)(nG - i)G^{i-1}(1 - G)^{n-i-1} dG, \tag{4.98}$$

which generalizes the formulas obtained earlier for the serial (4.26) and parallel (4.76) systems.

When the elements of the fault-tolerant system are jointly loaded by the stationary Poisson load flow, by analogy with the derivation from (4.86) of (4.88), using (4.98), we can obtain a generalized expression to predict the system's reliability function:

$$R_c(l, t, n) = \sum_{i=0}^{l-1} \binom{n}{i} \int_0^1 e^{-\omega_o t[1 - F_1(G)]}(nG - i)G^{i-1}(1 - G)^{n-i-1} dG. \tag{4.99}$$

In the expression (4.99), the level of the fault tolerance of the system structure is determined by the appropriate setting of a critical number of failures 1 in the range from 1 to n, and the used alternative of combining the distributions of the random load values and the limit load of the elements is provided by the choice of the function of the unit distribution $F_1(G)$, which depends on the safety factor (Table A.2).

Chapter 5
Prediction and Management of Reliability Under Conditions of Using Safety Devices

In many machine and aggregate designs, capabilities of the reliability management with safety factors can be limited because of variety of modes, operating conditions and functional features of elements, and the need to meet weight, economic and other criteria. Therefore, the method to prevent sudden failures by including special elements in the design—safeguards—has received a wide application in machines. Its main purpose is to regulate and limit the amount of loads acting under extreme conditions on the main (protected) elements. Common mechanical safeguards are safety couplings [57], valves, deformation limiters for elastic elements and other devices [58], limiting the magnitude of extreme loads perceived by the protected elements. The reason for using safeguards may be the need to prevent sudden failures in those elements of a combined loaded system that can lead to the failure of others or have other serious consequences.

When investigating the mechanical reliability of systems that incorporate safeguards, the key issue is quantitative dependencies between the parameters of safeguards, determining the magnitude of the maximum loads acting on the main elements and the reliability function of the protected object. Establishing such dependencies that are necessary for effective reliability management, it should be taken into account that for the majority of real mechanical safeguards, as a rule, there is a random scattering of the level limiting the maximum load ("actuation load" of safeguards) transmitted to the protected elements.

The analysis of the well-known designs with mechanical safeguards shows that the nature of the variability in the magnitude of the maximum load to make a safeguard work, protecting the main elements, can be of two kinds. For structures with the first-class safeguard (clutches with collapsible pins, strain limiters, safety valves), the response load is constant in time and does not depend on the magnitude or number of perceived extreme loads, as well as on all other factors. Therefore, the actuation load of such safeguards can be considered as a random variable fixed in time. Its degree of the initial dispersion is determined, for example, by dissipating the strength of the slutch drive pins or the stiffness of the valve springs.

O. Grynchenko, O. Alfyorov, *Mechanical Reliability*,
https://doi.org/10.1007/978-3-030-41564-8_5

The second type of safeguard structures (friction, claw or ball-type clutches) is characterized by the fact that the actuation load in time is unstable and can be changed accidentally at each extreme loading. Here the actuation load is determined by the amount of cohesion forces (usually frictional forces) between the elements and can depend on the dynamics of the loading process. When constructing reliability models, these features of real structures with safeguards must be taken into account [59, 60, 61].

5.1 Reliability Models When Using Safeguards with Deterministic Actuation Load

Let us consider the simplest case of the constructive scheme consisting of the basic element protected from overloads by the idealized safeguard with a constant deterministic value of the actuation load D = const. This corresponds to the case when the scattering of the actuation load of the first type of safeguards can be neglected. Such a safeguard transforms any continuous distribution of extreme loads with the distribution function $F(P_{\text{н}})$ into a piecewise continuous distribution [62]; its distribution function has the form

$$\widetilde{F}(P_{\text{H}}) = \begin{cases} \dfrac{F(P_{\text{H}})}{F(D)}, & \text{when } P_{\text{H}} < D; \\ 1, & \text{when } P_{\text{H}} \geq D. \end{cases} \tag{5.1}$$

Then, if $g(P_{\text{п}})$ and $G(P_{\text{п}})$ are the density and the limit load function of the main element, then using safeguards, its reliability function for a random m-fold extreme loading (Fig. 1.1b) will be determined by the expression:

$$R_m(D) = \int\limits_0^\infty \left[\widetilde{F}(P)\right]^m g(P)dP = \frac{\int\limits_0^D F^m(P)g(P)dP}{F^m(D)} + 1 - G(D). \tag{5.2}$$

Proceeding to the unit distributions, we can rewrite (5.2) in the following form

$$R_m(D) = \frac{\int\limits_0^{G(D)} [F_1(G)]^m dG}{F^m(D)} + 1 - G(D). \tag{5.3}$$

In view of the fact that the first summand in (5.3) when $m \to \infty$ tends to zero, the reliability function of an element with a safeguard is limited below by a value of $R_\infty(D) = 1 - G(D)$. That is ensured for any number of extreme loads and any distribution law of extreme loads.

If the extreme load and the limit load of the element have the same variation coefficients and are distributed according to the Weibull law, then

$$F(P_{\text{H}}) = 1 - \exp\left[-\left(\frac{P_{\text{H}}}{a_{\text{H}}}\right)^b\right];$$

$$G(P_{\Pi}) = 1 - \exp\left[-\left(\frac{P_{\Pi}}{a_{\Pi}}\right)^b\right]. \quad (5.4)$$

Then it follows from (5.3) that

$$R_m(D) = \frac{\frac{b}{a_{\Pi}} \int\limits_0^D \left\{1 - \exp\left[-\left(\frac{P}{a_{\text{H}}}\right)^b\right]\right\}^m \left(\frac{P}{a_{\Pi}}\right)^{b-1} \exp\left[-\left(\frac{P}{a_{\Pi}}\right)^b\right] dP}{\left\{1 - \exp\left[-\left(\frac{D}{a_{\text{H}}}\right)^b\right]\right\}^m} + \exp\left[-\left(\frac{D}{a_{\Pi}}\right)^b\right]. \quad (5.5)$$

In the particular case when $m = 1$, after integrating from (5.5), we obtain an expression for the reliability function when the element protected by safeguards is loaded for the first time:

$$R_1(D) = \frac{\frac{\overset{\Pi}{a}}{a_{\Pi}^b + a_{\text{H}}^b} \left(1 - \exp\left\{-\left[\left(\frac{D}{a_{\text{H}}}\right)^b + \left(\frac{D}{a_{\Pi}}\right)^b\right]\right\}\right)}{1 - \exp\left[-\left(\frac{D}{a_{\text{H}}}\right)^b\right]}. \quad (5.6)$$

We introduce a relative parameter characterizing the adjustment of the safeguard and determined by the ratio of its actuation load to the mean value of the external extreme load, which would act on the element if a safeguard was absent: $M = D/\overline{P}_{\text{H}}$. Here, the expression (5.6) takes the form:

$$R_1(M) = \frac{R_1\left\{1 - \exp\left[M^b \Gamma^b (1 + 1/b)/R_1\right]\right\}}{1 - \exp\left[-M^b \Gamma^b (1 + 1/b)\right]}, \quad (5.7)$$

where $R_1 = \overline{K}^b/(\overline{K}^b + 1)$—a reliability function of an element without safeguards at the first loading; $\overline{K} = \frac{\overline{P}_{\Pi}}{\overline{P}_{\text{H}}} = \frac{a_{\Pi}}{a_{\text{H}}}$—a safety factor of the element without using safeguards.

In the general case of multiple loading of an element protected by safeguards, the reliability function is conveniently determined on the basis of (5.3), using numerical integration and the following expression:

$$R_m(M) = \frac{\int_0^{G(M)} \left[1-(1-G)^{\overline{K}^b}\right]^m dG}{\left\{1-\exp\left[-M^b\Gamma^b(1+{}^1\!/_b)\right]\right\}^m} + \exp\left[-\frac{M^b\Gamma^b(1+{}^1\!/_b)}{\overline{K}^b}\right], \quad (5.8)$$

where $G(M) = 1 - \exp\left[-\frac{M^b\Gamma^b(1+{}^1\!/_b)}{\overline{K}^b}\right]$.

It follows from (5.8) that for any distribution laws of extreme loads and a limit load distribution function of the form (5.4), the lower limit of the reliability function when $m \to \infty$, is given by

$$R_\infty(M) = \exp\left[-\frac{M^b\Gamma^b(1+{}^1\!/_b)}{\overline{K}^b}\right]. \quad (5.9)$$

With the help of (5.9), it is easy to obtain an expression to calculate a value of safeguard generic parameter that guarantees a given reliability function of γ for any number of extreme loads

$$M_\gamma = \frac{\overline{K}(\ln{}^1\!/_\gamma)^{{}^1\!/_b}}{\Gamma(1+{}^1\!/_b)}. \quad (5.10)$$

Calculations of the reliability function, carried out with the help of expressions (5.7) and (5.8), show how the reliability function of the element protected by the safeguard increases with decreasing the generic parameter M. As an illustration of this, Fig. 5.1 shows the graphs (solid lines 1, 2 and 3) of the change in the reliability

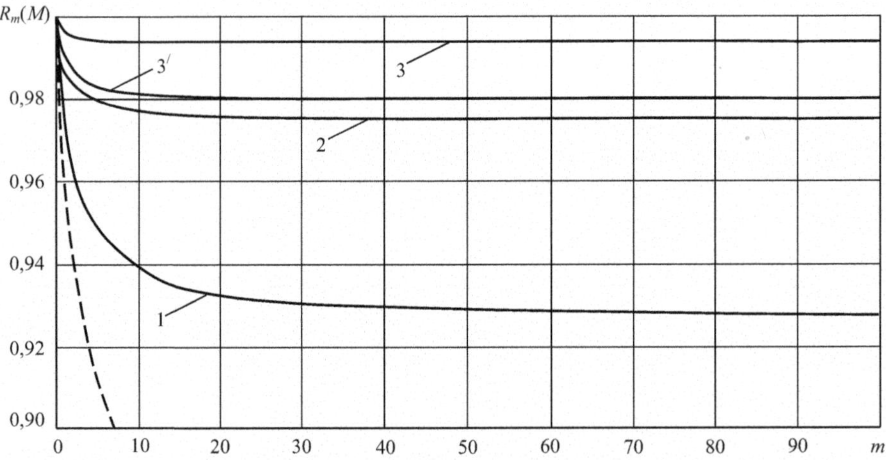

Fig. 5.1 Graphs of the dependencies of reliability function on the number of extreme loads: solid line, (*1*) element with safeguard, when $M = 1.1$; (2) element with safeguard, when $M = 1.0$; (*3*) element with safeguard, when $M = 0.9$; (*3'*) consistent system with safeguard when $n = 3$; $M = 0.9$; dashed line, element without safeguard

function of an element with a safeguard as a function of the number of loads, corresponding to decreasing values of the generic parameter M: 1.1; 1.0 and 0.9.

The graphs are constructed for the case when the safety factor $\overline{K} = 1.3$, and the load variation coefficients and limit load $V_{н} = V_{п} = 0.1$, which corresponds to the value of the shape parameter $b = 12.15$. The dashed line in Fig. 5.1 shows a graph of decreasing reliability function of an element not protected by the safeguard.

As can be seen from Fig. 5.1 the use of safeguards when $M = 0.9 \div 1.0$ in the case under consideration allows to sharply increase the reliability function of the element in comparison with the case when the overload protection is absent. It is significant that after a certain number ($m \approx 20$) of extreme loads, the level of reliability function is established at a value practically coinciding with its lower boundary, determined by the expression (5.9). When using safeguards, it should be taken into account that the possibility of increasing the reliability function only due to a decrease in the actuation load of the safeguard (parameter M) is limited in practice. This is due to the fact that, under normal operating conditions of the element loading, the probability of safeguard triggering should be small enough so that the technological process performed by the machine is not disturbed. Therefore, a rational level of a generic parameter can be selected when it is designed by controlling two quantities: M and $\overline{K}$. As follows from (5.10), the ratio $M_{\gamma}/\overline{K}$ depends only on the given reliability function γ. Therefore, it is possible to determine first the minimum level of the generic parameter $M_{\min}$, based on the permissible probability of the safeguard triggering in the operating mode of the element loading, and then calculate the value of the safety factor $\overline{K}_{\gamma}$ ensuring a given reliability function γ under extreme loading:

$$\overline{K}_{\gamma} = \frac{M_{\min}\Gamma(1 + 1/b)}{(\ln 1/\gamma)}. \tag{5.11}$$

For example, if the ordinate P_o of an ergodic stationary random process of loading an element in the operating mode is distributed in any section of the process according to the Weibull law with the distribution function

$$F(P_o) = 1 - \exp\left[-\left(\frac{P_o}{a_o}\right)^{b_o}\right],$$

then the probability that the workload exceeds a certain minimum level of the actuation load $D_{\min}$ is determined by the expression

$$Pr(P_0 > D_{min}) = exp\left[-\left(\frac{D_{min}\Gamma(1 + 1/b_0)}{\overline{P_0}}\right)\right]$$

where $\overline{P}_o$ is an average workload per element.

Denote α_∂ as a permissible value of this probability, which coincides with the probability of safeguard operation in the operating mode. Then the corresponding minimum level of the generic parameter of the safeguard is given by

$$M_{\min} = \frac{D_{\min}}{\overline{P}_{\text{H}}} = \frac{\overline{P}_o \left(\ln {}^1/_{\alpha_\partial}\right)^{1/b_o}}{\overline{P}_{\text{H}} \Gamma\left(1 + {}^1/_{b_o}\right)}.$$

This value can be used to determine the safety factor $\overline{K}_\gamma$ by the formula (5.11).

With a deterministic approach to the design calculation of safety clutches in the technical literature [57], it is recommended to regulate only the value of the ratio $D_{\min}/\overline{P}_o$, taking it equal to 1.25. If, for example, the variation coefficient of the load in the operating mode is $V_o = 0.15$, then based on the above expressions at this value of the ratio, the probability of the clutch actuation in the operating mode of loading of the protected element will be $\alpha_\partial = 0.0269 \approx 2.7\%$. However, in such a one-sided approach to the selection of the parameters of the safeguard, its effectiveness in fulfilling the main task—protecting the protected object from overloads—will remain undefined. Therefore, ensuring the reliability of the object, it is necessary to conform the level of the actuation load $D_{\min}$ with the safety factor of the element, calculated without taking into account the influence of the safeguard on extreme loads: $\overline{K} = \overline{P}_{\Pi}/\overline{P}_{\text{H}}$. It is also necessary to take into account the variation coefficient of the limit load of the element.

So, for example, if we assume that $D_{\min}/\overline{P}_o = 1.25$, then in the case when the average value of the extreme load is $\overline{P}_{\text{H}} = 1.5\overline{P}_o$, we obtain that $M_{\min} = 0.833$. Then, to guarantee the reliability function of the protected element $\gamma = 0.999$ from (5.11), it follows that for the variation coefficient of the limit load of the element $V_{\Pi} = 0.1$ it is also necessary that the safety factor $\overline{K}_{0.999}$ won't be less than 1.41.

If the load and limit load are distributed according to the log-logistic law with the same shape parameters and, therefore, the corresponding distribution functions have the form:

$$F(P_{\text{H}}) = \frac{\left(P_{\text{H}}/C_{\text{H}}\right)^\nu}{1 + \left(P_{\text{H}}/C_{\text{H}}\right)^\nu}; \quad G(P_{\Pi}) = \frac{\left(P_{\Pi}/C_{\Pi}\right)^\nu}{1 + \left(P_{\Pi}/C_{\Pi}\right)^\nu}.$$

From (5.2) we find that

$$R_m(D) = \frac{\nu \cdot \left[1 + \left(D/C_{\text{H}}\right)^\nu\right]^m}{C_{\Pi}\left(D/C_{\text{H}}\right)^{\nu m}} \int_0^D \frac{\left(P/C_{\text{H}}\right)^{\nu m} \cdot \left(P/C_{\Pi}\right)^{\nu - 1} dP}{\left[1 + \left(P/C_{\text{H}}\right)^\nu\right]^m \left[1 + \left(P/C_{\Pi}\right)^\nu\right]^2} + \frac{1}{1 + \left(D/C_{\Pi}\right)^\nu}. \tag{5.12}$$

When $m = 1$, then from (5.12) after integration, denoting $q_0 = \left(C_H/C_\Pi\right)^\nu$, we obtain an expression for the reliability function during the first extreme loading of an element with safeguards:

$$R_1(D) = \begin{cases} \dfrac{1}{1-q_o} + \dfrac{q_o\left[q_o + \left(D/C_\Pi\right)^\nu\right]}{(1-q_o)^2\left(D/C_\Pi\right)^\nu} \ln \dfrac{q_o\left[1 + \left(D/C_\Pi\right)^\nu\right]}{q_o + \left(D/C_\Pi\right)^\nu}; & \text{when } q_o \neq 1; \\ \dfrac{2 + \left(D/C_\Pi\right)^\nu}{2\left[1 + \left(D/C_\Pi\right)^\nu\right]}; \quad \text{when } q_o = 1. \end{cases} \tag{5.13}$$

In the general case, when $m > 1$ it is convenient to determine the reliability function, based on (5.3), using the numerical integration of the unit distribution function. After going to the relative generic parameter M, the corresponding expression takes the following form:

$$R_m(M) = \frac{\overline{K}^{\nu m}\left[1 + M^\nu\left(\frac{\pi}{\nu \sin \pi/\nu}\right)^\nu\right]^m}{M^{\nu m}\left(\frac{\pi}{\nu \sin \pi/\nu}\right)^{\nu m}} \int\limits_0^{G(M)} \frac{G^m dG}{\left[(1-G) + \overline{K}^\nu G\right]^m} + \frac{\overline{K}^\nu}{\overline{K}^\nu + M^\nu\left(\frac{\pi}{V \sin \pi/\nu}\right)^\nu}, \tag{5.14}$$

where $\overline{K} = \frac{C_\Pi}{C_H}$, $G(M) = \frac{M^\nu\left(\frac{\pi}{\nu \sin \pi/\nu}\right)^\nu}{\overline{K}^\nu + M^\nu\left(\frac{\pi}{V \sin \pi/\nu}\right)^\nu}$.

It follows from (5.14) that for any distribution law of extreme loads, if the limit load is distributed logarithmically and logically, the lower limit of the reliability function is determined from the expression:

$$R_\infty(M) \frac{\overline{K}^\nu}{\overline{K}^\nu + M^\nu\left(\frac{\pi}{\nu \sin \pi/\nu}\right)^\nu}, \tag{5.15}$$

and the formula to calculate the value of the safeguard generic parameter that guarantees a reliability function γ, has the following form

$$M_\gamma = \frac{\overline{K}\nu \sin \pi/\nu}{\pi}\left(\frac{1-\gamma}{\gamma}\right)^{1/\nu}. \tag{5.16}$$

A comparative analysis of the results of calculations carried out with the help of formulas (5.9) and (5.15) showed that under uncertainty conditions with respect to the type of the distribution law of the limit load in order to guarantee a high enough (lower limit) of the reliability function: $R_{\infty}(M) > 0,9$ When choosing values of the generic parameter M and the safety factor $\overline{K}$, the expressions (5.10) and (5.11) should be used.

For many real constructions with safeguards, it is advisable to treat the protected object as a system consisting of several elements. When building reliability models in this case, it is necessary to take into account the structure of the system and the number of its elements. The most common in machines are systems with a serial structure in the sense of reliability, when a failure of any element leads to a failure of the entire system. In the case of a combined loaded serial system consisting of n identical equally loaded elements and protected against overloads by a safeguard with a constant actuation load, using single distributions, by analogy with (5.3), we can obtain an expression for the reliability function in the following form:

$$R_{cm}(D) = \frac{n}{F^m(D)} \int_0^{G(D)} (1-G)^{n-1}[F_1(G)]^m dG + [1-G(D)]^n. \quad (5.17)$$

If the random values of the load on the system and the strength of its elements have a Weibull distribution with identical variation coefficients, then it follows from (5.17) that

$$R_{cm}(M) = \frac{n \int^{G(M)} (1-G)^{n-1}\left[1-(1-G)^{K^b}\right]^m dG}{\left\{1-\exp\left[-M^b\Gamma^b(1+1/b)\right\}^m + \exp\left[-\frac{nM^b\Gamma^b(1+1/b)}{\overline{K}^b}\right]}. \quad (5.18)$$

Figure 5.1 shows the results of calculating with the help of (5.18) the reliability function of a consistent system (curve 3′) consisting of three equally loaded elements for $\overline{K} = 1.3$; $V_{\text{н}} = V_{\text{п}} = 0.1$ and the value of the generic parameter of a safeguard: $M = 0.9$. This graph shows how much the reliability function of a serial system protected by a safeguard decreases, as compared to the reliability function of one of its elements (curve 3).

In case when the safeguard is intended to protect a jointly loaded system consisting of n equally loaded elements and having a parallel structure, an expression can be obtained to determine the reliability function of this system in the form:

$$\widetilde{R}_{cm}(D) = \frac{n}{F^m(D)} \int_0^{G(D)} G^{n-1}(F_1(G))^m dG + 1 - G^n(D). \quad (5.19)$$

The above reliability models of elements and systems protected from repeated impact of overloads by means of a safeguard with a nearly constant actuation load can serve as a theoretical basis when being designed and during modernization of machinery elements to manage reliability rationally.

5.2 Reliability Models when Using Safeguards with Random Actuation Load

Let us consider the case, when the actuation load of a safeguard depends only on the initial quality and it can be considered as a fixed in time random variable. In this case, its distribution density can serve as an comprehensive feature of the random scattering of the actuation load $\varphi(D)$. Then, based on the formula of the total probability, to determine the reliability function of an element protected by the first type of a safeguard, one can use the expression:

$$R_m = \int_0^\infty \varphi(D) R_m(D) dD. \tag{5.20}$$

Taking into account (5.2), the expression (5.20) can be represented in the form

$$R_m = \int_0^\infty \left[\frac{\varphi(D)}{F^m(D)} \int_0^D F^m(P) g(P) dP \right] dD + \int_0^\infty \varphi(D)[1 - G(D)] dD. \tag{5.21}$$

It follows from (5.21) that when $m \to \infty$, R_m is limited below and the lower bound of the reliability function of the element for any distribution law of extreme loads is given by

$$R_\infty = \int_0^\infty \varphi(D)[1 - G(D)] dD. \tag{5.22}$$

This means that the lower limit for the reliability function can be obtained as the reliability function of the element during the first loading by a random load, the distribution of which coincides with the distribution of the actuation load of the safeguard. Therefore, if the limit load of an element and the load of a safeguard operation are distributed according to the Weibull law with the same variation coefficients, then, using the analogy with (2.26), we find that

$$R_\infty = \frac{\overline{K}^b}{\overline{K}^b + \overline{M}^b}, \tag{5.23}$$

where $\overline{K} = \overline{P}_{\text{П}} / \overline{P}_{\text{H}}$—a safety factor;

$\overline{M} = \overline{D} / \overline{P}_{\text{H}}$—a reserve by a mean value of the generic parameter of a safeguard and a load;

$\overline{D}$—a mean value of the actuation load of the safeguard.

A comparison of the results of calculating the lower limit of the reliability function of an element with a safeguard using formulas (5.9) and (5.23) shows that taking into account the presence of scattering of the actuation load leads to a decrease in the value R_∞. The value of the reserve for the average generic parameter of a safeguard, guaranteeing a given reliability function of γ for any number of extreme loads, based on (5.23) is determined by the formula

$$\overline{M}_\gamma = \overline{K}\left(\frac{1-\gamma}{\gamma}\right)^{1/b}. \tag{5.24}$$

If the limit load of the element and the actuation load of the safeguard have a log-logistic distribution with identical variation coefficients, then using (2.79), it is possible to obtain for the lower limit of the reliability function the expression

$$R_\infty = \begin{cases} \dfrac{\overline{K}^\nu\left[\overline{K}^\nu - \nu\overline{M}^\nu\left(\ln\overline{K} - \ln\overline{M}\right) - \overline{M}^\nu\right]}{\left(\overline{K}^\nu - \overline{M}^\nu\right)^2}, & \text{when } \overline{M} \neq \overline{K}; \\ \dfrac{1}{2}, \quad \text{when } \overline{M} = \overline{K}. & \end{cases} \tag{5.25}$$

In the case where the variation coefficients of the actuation load of a safeguard and the limit load of the element are different, the calculated dependences for R_∞ and $\overline{M}_\gamma$ for the four laws of the load capacity distribution and the actuation load are given in the summary Table 5.1. It indicates: V_{M}—a variation coefficient of the actuation load of a safeguard: b_{M}—a shape parameter of the Weibull distribution of the actuation load: ν_{M}—a shape parameter of the log-logistic distribution of the actuation load:

In practice, difficulties in estimating the lower bound of the reliability function and the value of the average generic parameter of a safeguard that provides a given reliability function can arise because of the uncertainty regarding the form of the laws of distribution of the limit load of the protected object and the actuation load of the safeguard. In this case, for the guaranteed determination of values $\overline{M}_\gamma$, the values of the upper bounds for safety factors at the first loading $\overline{K}_{1\gamma}$ can be used. They are given in Table 3.3. The corresponding formula, which determines the lower limit of the average generic parameter of the safeguard, has the following form:

$$\overline{M}_\gamma = \overline{K}/\overline{K}_{1\gamma}, \tag{5.26}$$

where $\overline{K}_{1\gamma}$ is selected from Table 3.3 in accordance with the required lower limit for the reliability function γ and the variation coefficient of the actuation load $V_{\text{H}} = V_{\text{M}}$.

So, for instance, if for any number of extreme loads, the reliability function of an element with a safeguard should not be less than 0.99, and the variation coefficient of variation V_{M} of the actuation load of a safeguard is 0.2, then for the variation coefficient of the limit load of the protected element 0.1 we determine the

Table 5.1 Calculation dependence for R_∞ and $\overline{M}_\gamma$

	Distribution laws of limit load and load of safeguards	Dependencies to calculate the lower limit of reliability function and generic parameter of safeguards
1	2	3
1	Normal	$R_\infty = F_o\left(\frac{\overline{K}-\overline{M}}{\sqrt{V_{\Pi}^2\overline{K}^2+V_{M}^2\overline{M}^2}}\right)$ $\overline{M}_\gamma = \frac{\overline{K}\left(1-U_\gamma^2V_{\Pi}^2\right)}{1+U_\gamma\sqrt{V_{\Pi}^2+V_{M}^2-\left(U_\gamma V_{\Pi}V_{M}\right)^2}}$
2	Log-normal	$R_\infty = F_o\left(\frac{\ln\overline{K}-\ln\overline{M}-\ln\sqrt{\frac{1+V_{\Pi}^2}{1+V_{M}^2}}}{\sqrt{\ln\left[\left(1+V_{\Pi}^2\right)\cdot\left(1+V_{M}^2\right)\right]}}\right)$ $\overline{M}_\gamma = \frac{\overline{K}\sqrt{1+V_{M}^2}}{\sqrt{1+V_{\Pi}^2}\cdot\exp\left(U_\gamma\sqrt{\ln\left[\left(1-V_{\Pi}^2\right)\left(1+V_{M}^2\right)\right]}\right)}$
1	2	3
3	Weibull law	$R_\infty = 1-\int_0^1 \exp\left(-\eta\left[\ln{}^1/_{(1-x)}\right]^\beta\right)dx,$ At $\beta = {}^{b_M}/_{b_\Pi}$; $\eta = \left(\frac{\overline{K}\Gamma\left(1+{}^1/_{b_M}\right)}{\overline{M}\Gamma\left(1+{}^1/_{b_\Pi}\right)}\right)^{b_M}$ At $b_M > 0.3b_\Pi$ and $\gamma \geq 0.99$ $R_\infty = 1-\left(\frac{\overline{M}\ \Gamma\left(1+{}^1/_{b_\Pi}\right)}{\overline{K}\Gamma\left(1+{}^1/_{b_M}\right)}\right)^{b_\Pi}\cdot\Gamma\left(1+{}^{b_\Pi}/_{b_M}\right)$ $\overline{M}_\gamma = K\frac{\Gamma\left(1+{}^1/_{b_M}\right)}{\Gamma\left(1+{}^1/_{b_\Pi}\right)}\left(\frac{1-\gamma}{\Gamma\left(1+{}^b/_{b_M}\right)}\right)^{{}^1/_{b_\Pi}}$
4	Log-logistic	$R_\infty = \int_0^1 \frac{x^\delta dx}{q(1-x)^\delta + x^\delta},$ at $\delta = \frac{\nu_M}{\nu_\Pi}$; $q = \left(\frac{\overline{M}\delta\sin\frac{\pi}{\nu_M}}{\overline{K}\sin\frac{\delta\pi}{\nu_M}}\right)^{\nu_M}$

corresponding value $\overline{K}_{1\gamma} = 1.82$ from Table 3.3. Then, if a safety factor of the element without using safeguards: $\overline{K} = 1.3$, then to ensure a given level of reliability it is necessary that the guaranteed mean value of the relative generic parameter of the safeguard is not more than $\overline{M}_{0.99} = {}^{1.3}/_{1.82} = 0.714$. If it is required to provide a higher lower "threshold" of the reliability function: $\gamma = 0.999$, then it follows from the data of Table 3.3 that for this purpose it is necessary to reduce the average generic parameter of a safeguard to the value of $\overline{M}_{0.999} = {}^{1.3}/_{2.511} = 0.517$. To increase the average generic parameter of the safeguard $\overline{M}_{0.999}$ without changing the reliability function is possible by decreasing scattering of the safeguard of the actuation load. So, if we ensure that $V_M = 0.1$, then under the conditions of the example using Table 3.3 we find that $\overline{M}_{0.999} = {}^{1.3}/_{1.863} = 0.698$.

In conclusion, we will consider a safeguard with a time-varying random actuation load. Friction clutches and some other types of safeguards do not provide stability in time of a random actuation load. Schematizing the external extreme loading of an object protected by such a safeguard, we will proceed, as before, from a continuous random process to a discrete stream of stochastically independent random effects (Fig. 1.1).

Similarly, when constructing a theoretical model of the reliability of an object protected by a safeguard of this type from overloads, a scheme will be considered in accordance with which, for each external extreme action, the loading of the safeguard is a random variable. Its realization is stochastically independent of its previous values and the value of the external extreme load which would act on the object in the absence of a safeguard.

We will assume that the distribution function $H(P)$ of the safeguard actuation load is known, when the power flow loading the protected object is interrupted. In this case, the acting external extreme load having the distribution function $F(P)$ is subjected to an independent random censoring [37] with the help of the safeguard, as a result of which its distribution function is transformed. Let us denote $\widetilde{F}(P)$ the distribution function of the load acting on the object after censoring with the help of a safeguard. Such a function must satisfy an equation of the form:

$$(1 - F(P))(1 - H(P)) = 1 - \widetilde{F}(P).$$

Therefore, the load distribution function acting on an object protected by the safeguard is determined using the expression

$$\widetilde{F}(P) = F(P) + H(P) - F(P)H(P). \tag{5.27}$$

In estimating the reliability function of elements and systems protected by a safeguard, it is convenient to pass to (5.27) [61] to unit distributions of the actuation load $H_1(G)$ and external load $F_1(G)$, after which

$$\widetilde{F}_1(G) = F_1(G) + H_1(G) - F_1(G)H_1(G). \tag{5.28}$$

Then the reliability function of an object protected by a safeguard under m-fold extreme loading is determined from an expression of the form:

$$R(m) = \int_0^1 \{F_1(G) + H_1(G) - F_1(G)H_1(G)\}^m dG. \tag{5.29}$$

Let us consider an example of constructing the simplest model to predict the reliability of an element, protected by a safeguard, on the basis of the general expression (5.29). Suppose that the extreme load, the limit load of the protected object and the actuation load of the safeguard are distributed according to the Frechet law (1.8) with the same level of relative scattering determined by the shape parameter ρ. Then the functions of the unit distributions of the external extreme load $F_1(G)$ and the actuation load $H_1(G)$ have the form:

$$F_1(G) = G^{1/\overline{K}^\rho};\ H_1(G) = G^{1/\overline{L}^\rho}, \tag{5.30}$$

where $\overline{K} = \frac{\overline{P}_\Pi}{\overline{P}_H}$—a safety factor of the protected element;

$\overline{L} = \frac{\overline{P}_\Pi}{\overline{P}_M}$—a relative level of the safeguard adjustment, calculated from the average values of the limit load $\overline{P}_\Pi$ and actuation load $\overline{P}_M$.

Substituting (5.30) into (5.29), we obtain an expression to predict reliability function of an element with a safeguard under m-fold loading:

$$R(m) = \int_0^1 \left\{ G^{1/\overline{K}^\rho} + G^{1/\overline{L}^\rho} - G^{1/\overline{K}^\rho + 1/\overline{L}^\rho} \right\}^m dG. \tag{5.31}$$

The first loading from (5.31) yields a simple analytic expression to predict the reliability function of an element using a safeguard:

$$R(1) = \frac{\overline{K}^\rho}{\overline{K}^\rho + 1} + \frac{\overline{L}^\rho}{\overline{L}^\rho + 1} - \frac{\overline{K}^\rho \overline{L}^\rho}{\overline{K}^\rho + \overline{L}^\rho + \overline{K}^\rho \overline{L}^\rho}. \tag{5.32}$$

The first summand of the sum (5.32) predicts the reliability of the element without using safeguards (when $L = 0$). The remaining terms allow you to control the reliability function by increasing the generic parameter L. So, for example, if the element has a safety factor, and the variation coefficient $\overline{K} = 1.3$ of the extreme load and limit load $V_H = V_\Pi = 0.1$, which corresponds $\rho = 13.62$, then its reliability function without the use of a safeguard: $\frac{1.3^{13.62}}{1.3^{13.62}+1} = 0.9727$. Then, using a safeguard with the variation coefficient of the actuation load $V_M = 0.1$ at the value of the generic parameter $\overline{L} = 1.2$ with the help of (5.32), we get that the predicted reliability function should rise to a value of $R(1) = 0.996$.

Appendix

See Tables A.1, A.2, A.3, and A.4.

O. Grynchenko, O. Alfyorov, *Mechanical Reliability*,
https://doi.org/10.1007/978-3-030-41564-8

Table A.1

γ	K										
	1.5	1.6	1.7	1.8	1.9	2.0	2.1	2.2	2.3	2.4	2.5
$V_п = 0,\ V_н = 0.08$											
0.9	180.08	532.78	1476	3858	9574	22,675	51,494	112,557	237,626	485,960	965,209
0.95	87.67	259.38	718.61	1878	4661	11,039	25,069	54,797	115,685	236,583	469,898
0.99	17.18	50.82	140.80	368.04	913.28	2163	4912.01	10,736	22,667	46,355	92,071
0.999	1.71	5.06	14.02	36.64	90.92	215.33	488.99	1068.85	2256.50	4614.67	9165.60
$V_п = 0,\ V_н = 0.1$											
0.9	50.05	120.47	275.02	598	1250	2515	4888	9212	16,877	30,133	52,543
0.95	24.37	58.65	133.89	291.61	608	1224	2380	4484	8216	14,670	25,579
0.99	4.77	11.49	26.23	57.14	119.32	239.95	466.34	878.77	1609.94	2874.44	5012
0.999	0.48	1.14	2.61	5.69	11.87	23.89	46.42	87.48	160.27	286.15	498.95
$V_п = 0,\ V_н = 0.12$											
0.9	21.48	45.05	90.41	174	324	585	1026	1752	2922	4767	7623
0.95	10.45	21.93	44.02	84.91	158.10	285	499	853	1422	2320	3711
0.99	2.05	4.30	8.62	16.64	30.98	55.87	97.92	167.18	278.74	454.73	727
0.999	0.20	0.43	0.86	1.66	3.08	5.56	9.75	16.64	27.75	45.27	72.39
$V_п = 0,\ V_н = 0.2$											
0.9	4.08	6.48	10.04	15.17	22.44	32.54	46.35	64.95	89.66	122.10	164.21
0.95	1.98	3.16	4.89	7.39	10.92	15.84	22.56	31.62	43.65	59.44	79.94
0.99	0.39	0.62	0.96	1,,4471	2.14	3.10	4.42	6.1951	8.55	11.65	15.66
0.999	0.04	0.06	0.10	0.14	0.21	0.31	0.44	0.62	0.85	1.16	1.56

Table A.2

Weibull distribution	Distribution function of load $F(P) = 1 - \exp\left[-\left(\frac{P}{a_{H}}\right)^{b_{H}}\right].$ Distribution function of limit load $G(P) = 1 - \exp\left[-\left(\frac{P}{a_{\Pi}}\right)^{b_{\Pi}}\right].$	Unit distribution function of load $F_1(G) = 1 - \exp\left\{-\eta(\overline{K})\left(\ln\frac{1}{1-G}\right)^{\beta}\right\},$ where $\eta(\overline{K}) = \overline{K}^{b_{H}}\left(\frac{\Gamma(1+1/b_{H})}{\Gamma(1+1/b_{\Pi})}\right)^{b_{H}};$ $\beta = \frac{b_{H}}{b_{\Pi}}.$ Unit distribution function of limit load $G_1(F) = 1 - \exp\left\{-\left[\eta(\overline{K})\right]^{-1/\beta}\left(\ln\frac{1}{1-F}\right)^{1/\beta}\right\}.$ At $b_{H} = b_{\Pi} = b$ $F_1(G) = 1 - (1-G)^{\overline{K}^{b}};$ $G_1(F) = 1 - (1-F)^{1/\overline{K}^{b}}.$
Frechet distribution	Distribution function of load $F(P) = \exp\left[-\left(\frac{h_{H}}{P}\right)^{\rho_{H}}\right].$ Distribution function of limit load $G(P) = \exp\left[-\left(\frac{h_{\Pi}}{P}\right)^{\rho_{\Pi}}\right].$	Unit distribution function of load $F_1(G) = \exp\left\{-\mu(\overline{K})\left(\ln\frac{1}{G}\right)^{\alpha}\right\},$ where $\mu(\overline{K}) = \left(\frac{\Gamma(1-1/\rho_{\Pi})}{\overline{K}\Gamma(1-1/\rho_{\Pi})}\right)^{\rho_{\Pi}};$ $\alpha = \frac{\rho_{H}}{\rho_{\Pi}}.$ Unit distribution function of limit load $G_1(F) = \exp\left\{-\left[\mu(\overline{K})\right]^{-1/\alpha}\left(\ln\frac{1}{F}\right)^{1/\alpha}\right\}.$ At $\rho_{H} = \rho_{\Pi} = \rho$ $F_1(G) = G^{1/\overline{K}^{\rho}};$ $G_1(F) = F^{\overline{K}^{\rho}}.$
Log-logistic distribution	Distribution function of load $F(P) = \frac{(P/C_{H})^{\nu_{H}}}{1+(P/C_{H})^{\nu_{H}}}.$ Distribution function of limit load $G(P) = \frac{(P/C_{\Pi})^{\nu_{\Pi}}}{1+(P/C_{\Pi})^{\nu_{\Pi}}}.$	Unit distribution function of load $F_1(G) = \frac{G^{\delta}}{q(\overline{K})(1-G)^{\delta}+G^{\delta}},$ where $q(\overline{K}) = \left(\frac{\nu_{H}\sin\frac{\pi}{\nu_{H}}}{\overline{K}\nu_{\Pi}\sin\frac{\pi}{\nu_{\Pi}}}\right)^{\nu_{H}};$ $\delta = \frac{\nu_{H}}{\nu_{\Pi}}.$ Unit distribution function of limit load $G_1(F) = \frac{\left[q(\overline{K})\right]^{1/\delta}F^{1/\delta}}{\left[q(\overline{K})\right]^{1/\delta}F^{1/\delta}+(1-F)^{1/\delta}}.$ At $\nu_{H} = \nu_{\Pi} = \nu$ $F_1(G) = \frac{\overline{K}^{\nu}G}{1+G(\overline{K}^{\nu}-1)};$ $G_1(F) = \frac{F}{F+\overline{K}^{\nu}(1-F)}.$
Combination of Frechet distribution and log-logistic distribution	Distribution function of load $F(P) = \exp\left[-\left(\frac{h}{P}\right)^{\rho}\right].$ Distribution function of limit load $G(P) = \frac{(P/C)^{\nu}}{1+(P/C)^{\nu}}.$	Unit distribution function of load $F_1(G) = \exp\left[-A(\overline{K})\frac{(1-G)^{\rho/\nu}}{G^{\rho/\nu}}\right],$ where $A(\overline{K}) = \left(\frac{\rho\sin\frac{\pi}{\rho}\Gamma(1+1/b)}{\overline{K}\;\nu\;\sin\frac{\pi}{\nu}}\right)^{\rho};$ Unit distribution function of limit load $G_1(F) = \frac{\left(A(\overline{K})\right)^{\nu/\rho}}{\left(A(\overline{K})\right)^{\nu/\rho}+\left(\ln\frac{1}{F}\right)^{\nu/\rho}}.$ At $\rho = \nu$ $F_1(G) = \exp\left[-\left(\frac{\Gamma(1+1/b)}{\overline{K}}\right)^{\rho}\frac{(1-G)}{G}\right];$ $G_1(F) = \frac{\Gamma^{\rho}(1+1/b)}{\Gamma^{\rho}(1+1/b)-\overline{K}^{\rho}\ln F}.$

(continued)

Combination of Frechet distribution and Weibull distribution	Distribution function of load $F(P) = \exp\left[-\left(\frac{h}{P}\right)^{\rho}\right]$. Distribution function of limit load $G(P) = 1 - \exp\left[-\left(\frac{P}{a}\right)^{b}\right]$.	Unit distribution function of load $F_1(G) = \exp\left\{-\theta(\overline{K})\left[-\ln(1-G)^{-\rho/b}\right]\right\}$, where $\theta(\overline{K}) = \left(\frac{\Gamma(1+1/b)}{\overline{K}\Gamma(1+1/b)}\right)^{\rho}$. Unit distribution function of limit load $G_1(F) = 1 - \exp\left[-\left(\theta(\overline{K})\right)^{b/\rho}\left(\ln\frac{1}{F}\right)^{b/\rho}\right]$. At $\rho = b$ $F_1(G) = \exp\left\{-\widetilde{\theta}(\overline{K})\left[-\ln(1-G)\right]^{-1}\right\}$ $G_1(F) = 1 - \exp\left[\frac{\widetilde{\theta}(\overline{K})}{\ln F}\right]$, where $\widetilde{\theta}(\overline{K}) = \frac{\Gamma^{2p}(1+1/\rho)\left(\rho\sin\frac{\pi}{\rho}\right)^{\rho}}{\left(\pi\overline{K}\right)^{\rho}}$.

Table A.3

γ	$\overline{K}$										
	1.5	1.6	1.7	1.8	1.9	2	2.1	2.2	2.3	2.4	2.5
$V_{п} = 0.08$, $V_{н} = 0.08$											
0.9	74.19	219.14	606.85	1586	3935	9320	21,165	46,260	97,670	199,730	396,700
0.95	28.33	83.5	231.1	603.9	1498	3548	8057	17,610	37,180	76,040	151,020
0.99	3.64	10.53	28.95	75.5	187.3	443.6	1006	2199	4645	9500	18,870
0.999	0.285	0.73	1.86	4.7	11.5	27.2	61.1	133.8	283.8	576	1150
$V_{п} = 0.08$, $V_{н} = 0.1$											
0.9	26.755	64.21	146.43	318.75	665.5	1338.2	2600.5	4900.5	8978	16,028	27,950
0.95	11.252	26.925	61.33	133.45	278.55	560	1088.5	2051	3757	6708	11,697
0.99	1.782	4.175	9.44	20.47	42.7	85.8	166.6	313.9	575	1027	1790
0.999	0.165	0.381	0.86	1.85	3.85	7.7	14.95	28.2	51.8	92.5	161.5
$V_{п} = 0.08$, $V_{н} = 0.12$											
0.9	13.476	28.155	56.4	108.71	202.33	364.85	639.3	1091.5	1819.7	2968.5	4747
0.95	5.979	12.453	24.91	47.98	89.27	160.93	282	481.4	802.5	1309.3	2093.5
0.99	1.045	2.143	4.255	8.16	15.15	27.3	47.85	81.7	136.2	222	355
0.999	0.101	0.206	0.409	0.79	1.46	2.63	4.6	7.9	13.1	21.3	34.1
$V_{п} = 0.08$, $V_{н} = 0.2$											
0.9	3.359	5.3135	8.203	12.377	18.285	26.495	37.72	52.84	72.93	99.31	133.53
0.95	1.5985	2.525	3.894	5.871	8.67	12.56	17.876	25.043	34.56	47.05	63.27
0.99	0.307	0.484	0.745	1.122	1.655	2.397	3.412	4.775	6.59	8.97	12.06
0.999	0.0305	0.048	0.074	0.112	0.165	0.239	0.339	0.475	0.655	0.89	1.195
$V_{п} = 0.1$, $V_{н} = 0.08$											
0.9	49.51	146.05	404.2	1056.2	2620.5	6207	14,095	30,805	65,035	133,000	264,170
0.95	16.28	47.76	131.95	344.6	854.8	2024.5	4597	10,049	21,210	43,380	86,150
0.99	1.638	4.365	11.75	30.45	75.3	178	404	883	1865	3810	7570

(continued)

γ	$\overline{K}$										
	1.5	1.6	1.7	1.8	1.9	2	2.1	2.2	2.3	2.4	2.5
0.999	0.13	0.285	0.61	1.32	2.93	6.7	15	33	69	142	280
$V_п = 0.1$, $V_н = 0.1$											
0.9	19.78	47.359	107.88	234.75	490.05	985.3	1914.7	3608	6610	11,801	20,578
0.95	7.525	17.885	40.63	88.32	184.25	370.35	719.7	1356	2484	4435	7733
0.99	1.038	2.29	5.04	10.82	22.47	45.05	87.45	164.7	301.5	538.5	939
0.999	0.093	0.188	0.376	0.73	1.42	2.74	5.24	9.76	17.82	31.7	54.9
$V_п = 0.1$, $V_н = 0.12$											
0.9	10.654	22.185	44.38	85.47	159.02	286.7	502.35	857.6	1429.7	2332.5	3729.5
0.95	4.415	9.117	18.173	34.94	64.96	117.05	205.05	350	583.5	951.9	1522
0.99	0.711	1.399	2.71	5.14	9.5	17.06	29.85	50.9	84.8	138.3	221
0.999	0.0665	0.127	0.237	0.437	0.789	1.41	2.44	4.15	6.9	11.3	18
$V_п = 0.1$, $V_н = 0.2$											
0.9	3.0165	4.752	7.318	11.025	16.273	23.565	33.535	46.965	64.81	88.24	118.64
0.95	1.4097	2.211	3.395	5.106	7.527	10.892	15.495	21.69	29.93	40.74	54.77
0.99	0.2663	0.4148	0.633	0.948	1.393	2.01	2.855	3.99	5.5	7.485	10.06
0.999	0.0263	0.041	0.0625	0.0935	0.138	0.199	0.283	0.395	0.545	0.739	1

Table A.4

m	Ω									
	0.001	0.002	0.003	0.004	0.005	0.006	0.007	0.008	0.009	0.01
0	1	1	1	1	1	1	1	1	1	1
0.1	0.99985	0.99969	0.99954	0.99939	0.99924	0.99908	0.99893	0.99878	0.99863	0.99848
0.2	0.99971	0.99942	0.99914	0.99885	0.99856	0.99828	0.99799	0.99771	0.99742	0.99714
0.3	0.99959	0.99919	0.99878	0.99837	0.99797	0.99756	0.99716	0.99676	0.99635	0.99595
0.4	0.99948	0.99897	0.99846	0.99794	0.99743	0.99692	0.99641	0.99590	0.99539	0.99489
0.5	0.99939	0.99877	0.99816	0.99755	0.99694	0.99634	0.99573	0.99512	0.99452	0.99392
0.6	0.99939	0.99877	0.99816	0.99755	0.99694	0.99634	0.99573	0.99512	0.99452	0.99392
0.7	0.99921	0.99843	0.99765	0.99687	0.99609	0.99531	0.99454	0.99376	0.99299	0.99222
0.8	0.99914	0.99828	0.99742	0.99656	0.99571	0.99486	0.99400	0.99315	0.99321	0.99146
0.9	0.99907	0.99814	0.99721	0.99628	0.99536	0.99443	0.99351	0.99259	0.99167	0.99076
1	0.99900	0.99800	0.99701	0.99602	0.99502	0.99404	0.99305	0.99206	0.99108	0.99010
1.1	0.99894	0.99788	0.99682	0.99577	0.99471	0.99366	0.99261	0.99157	0.99052	0.98948
1.2	0.99888	0.99776	0.99665	0.99553	0.99442	0.99331	0.99221	0.99110	0.99000	0.98890
1.3	0.99882	0.99765	0.99648	0.99531	0.99414	0.99298	0.99182	0.99066	0.98950	0.98835
1.4	0.99877	0.99755	0.99632	0.99510	0.99388	0.99267	0.99145	0.99024	0.98904	0.98783
1.5	0.99872	0.99744	0.99617	0.99490	0.99363	0.99237	0.99111	0.98985	0.98859	0.98733
1.6	0.99867	0.99735	0.99603	0.99471	0.99340	0.99208	0.99077	0.98947	0.98816	0.98686
1.7	0.99863	0.99726	0.99589	0.99453	0.99317	0.99181	0.99046	0.98911	0.98776	0.98641
1.8	0.99858	0.99717	0.99576	0.99435	0.99295	0.99155	0.99015	0.98876	0.98737	0.98598
1.9	0.99854	0.99709	0.99564	0.99419	0.99274	0.99130	0.98986	0.98843	0.98700	0.98557
2	0.99850	0.99701	0.99552	0.99403	0.99254	0.99106	0.98959	0.98811	0.98664	0.98517

(continued)

m	Ω									
	0.001	0.002	0.003	0.004	0.005	0.006	0.007	0.008	0.009	0.01
2.1	0.99846	0.99693	0.99540	0.99387	0.99235	0.99083	0.98932	0.98781	0.98630	0.98479
2.2	0.99843	0.99686	0.99529	0.99373	0.99217	0.99061	0.98906	0.98751	0.98579	0.98443
2.3	0.99839	0.99678	0.99518	0.99358	0.99199	0.99040	0.98881	0.98723	0.98565	0.98407
2.4	0.99836	0.99671	0.99508	0.99344	0.99182	0.99019	0.98857	0.98695	0.98534	0.98373
2.5	0.99832	0.99665	0.99498	0.99331	0.99165	0.98999	0.98834	0.98669	0.98504	0.98340
2.6	0.99829	0.99658	0.99488	0.99318	0.99149	0.98980	0.98811	0.98643	0.98475	0.98308
2.7	0.99826	0.99652	0.99479	0.99306	0.99133	0.98961	0.98790	0.98618	0.98448	0.98277
2.8	0.99823	0.99646	0.99470	0.99294	0.99118	0.98943	0.98768	0.98594	0.98421	0.98247
2.9	0.99820	0.99640	0.99461	0.99282	0.99103	0.98926	0.98748	0.98571	0.98394	0.98218
3	0.99817	0.99634	0.99452	0.99270	0.99089	0.98908	0.98728	0.98548	0.98369	0.98190

m	Ω								
	0.02	0.03	0.04	0.05	0.06	0.07	0.08	0.09	0.1
0	1	1	1	1	1	1	1	1	1
0.1	0.9970	0.9955	0.9940	0.9926	0.9912	0.9898	0.9884	0.9871	0.9857
0.2	0.9943	0.9916	0.9888	0.9861	0.9835	0.9809	0.9783	0.9758	0.9733
0.3	0.9920	0.9881	0.9842	0.9804	0.9767	0.9730	0.9694	0.9658	0.9623
0.4	0.9899	0.9849	0.9800	0.9753	0.9706	0.9659	0.9614	0.9569	0.9525
0.5	0.9879	0.9821	0.9763	0.9706	0.9650	0.9595	0.9541	0.9488	0.9436
0.6	0.9862	0.9795	0.9729	0.9664	0.9600	0.9537	0.9475	0.9415	0.9355
0.7	0.9846	0.9771	0.9697	0.9625	0.9554	0.9484	0.9415	0.9348	0.9281
0.8	0.9831	0.9749	0.9668	0.9589	0.9511	0.9434	0.9359	0.9285	0.9213
0.9	0.9817	0.9728	0.9641	0.9555	0.9471	0.9389	0.9307	0.9228	0.9150
1	0.9804	0.9709	0.9615	0.9524	0.9434	0.9346	0.9259	0.9174	0.9091
1.1	0.9792	0.9691	0.9592	0.9495	0.9399	0.9306	0.9214	0.9124	0.9036
1.2	0.9780	0.9674	0.9569	0.9467	0.9367	0.9268	0.9172	0.9077	0.8984
1.3	0.9769	0.9658	0.9548	0.9441	0.9336	0.9233	0.9132	0.9033	0.8936
1.4	0.9759	0.9643	0.9528	0.9416	0.9307	0.9199	0.9094	0.8991	0.8890
1.5	0.9749	0.9628	0.9509	0.9393	0.9279	0.9168	0.9058	0.8951	0.8846
1.6	0.9740	0.9614	0.9491	0.9371	0.9253	0.9137	0.9024	0.8913	0.8805
1.7	0.9731	0.9601	0.9474	0.9350	0.9228	0.9109	0.98992	0.8878	0.8766
1.8	0.9723	0.9589	0.9458	0.9329	0.9204	0.9081	0.8961	0.8843	0.8728
1.9	0.9715	0.9577	0.9442	0.9310	0.9181	0.9055	0.8931	0.8811	0.8692
2	0.9707	0.9565	0.9427	0.9292	0.9159	0.9030	0.8903	0.8779	0.8658
2.1	0.9699	0.9554	0.9412	0.9274	0.9138	0.9006	0.8876	0.8749	0.8625
2.2	0.9692	0.9544	0.9398	0.9275	0.9118	0.8982	0.8850	0.8720	0.8594
2.3	0.9685	0.9533	0.9385	0.9240	0.9098	0.8960	0.8825	0.8693	0.8563
2.4	0.9679	0.9523	0.9372	0.9224	0.9080	0.8939	0.8801	0.8666	0.8534

(continued)

m	Ω							
	0.02	0.03	0.04	0.05	0.06	0.07	0.08	0.09	0.1
2.5	0.9672	0.9514	0.9360	0.9209	0.9062	0.8918	0.8777	0.8640	0.8506
2.6	0.9666	0.9505	0.9347	0.9194	0.9044	0.8898	0.8755	0.8615	0.8479
2.7	0.9660	0.9496	0.9336	0.9180	0.9027	0.8878	0.8733	0.8591	0.8453
2.8	0.9654	0.9487	0.9324	0.9166	0.9011	0.8860	0.8712	0.8568	0.8427
2.9	0.9648	0.9479	0.9313	0.9152	0.8995	0.8841	0.8692	0.8545	0.8403
3	0.9643	0.9471	0.9303	0.9139	0.8980	0.8824	0.8672	0.8524	0.8379

Literature

1. D.N. Reshetov, *Machine Elements* (Mashinostroenie, Moscow, 1989). 496 p
2. L.A. Sosnovsky, Calculation of details taking into account the probabilistic nature of strength and loading. Vestnik Mashynostroeniia (8), 8–10 (1974)
3. G.S. Pisarenko, A.A. Lebedev, *Deformation and Strength of Materials in Complex Stress state* (Naukova Dumka, Kiev, 1976). 415 p
4. A.O. Lebedev, M.I. Bobir, V.P. Lamashevskiy, *Materials Mechanics for Engineers* (NTU of Ukraine "KPI", Kiyv, 2006). 288 p
5. V.A. Svetlitskiy, *Random Oscillations of Mechanical Systems* (Mashinostroenie, Moscow, 1976). 216 p
6. O.V. Berestnev, Y.L. Soliterman, A.M. Goman, *Standardization of Reliability of Technical Systems: Monograph* (UP "Technoprint", Minsk, 2004). 266 p
7. R.V. Kugel, *Foreign Cars "Automobile Industry". Results of Science and Technology* (VINITI AN SSSR, Moscow, 1987). 156 p
8. S.A. Laptev, *Comprehensive Vehicle Test System: Formation, Development, Standardization* (Izdatelstvo Standartov, Moscow, 1991). 172 p
9. A.S. Gusev, in *Probabilistic Methods in Mechanics of Machines and Constructions*, ed. by A. S. Gusev, V. A. Svetlitskiy, (Publishing House MSTU Named After N.E. Bauman, Moscow, 2009). 224 p
10. I.V. Dunin-Barkovsky, N.V. Smirnov, *Probability Theory and Mathematical Statistics in Engineering* (Moscow, Gostechiaztad, 1955). 556 p
11. D.R. Cox, D. Ouks, *Analysis of Life-Time Type Data* (Finance and Statistics, Moscow, 1988). 192 p
12. G. Upton, I. Cook, *Oxford Dictionary of Statistics* (Oxford University Press, Oxford, 2008). 453 p
13. E. Gumbel, *Statistics of Extreme Values* (Mir, Moscow, 1965). 452 p
14. O.S. Grynchenko, Models for predicting strength of reliability of machine elements for a single fracture. Sci. Bull. KhNTU, Kharkiv **4**, 21–27 (2000)
15. N.L. Johnson, in *One-Dimensional Discrete Distributions*, ed. by N. L. Johnson, S. Kots, A. U. Kemp, (Binom, Moscow, 2012). 559 p
16. J. Bogdanoff, F. Kozin, *Probabilistic Models of Damage Accumulation* (Mir, Moscow, 1989). 344 p
17. N.L. Johnson, in *One-Dimensional Continuous Distributions*, ed. by N. L. Jonson, S. Kots, N. Balakrishnan, (Binom, Moscow, 2010). Part 1, 703 p
18. O.S. Grynchenko, Numerical solution of the problems of forecasting and providing strength reliability under multiple overloads. Bull. KhSTU Agric. Kharkiv **8**(1), 257–264 (2001)

O. Grynchenko, O. Alfyorov, *Mechanical Reliability*,
https://doi.org/10.1007/978-3-030-41564-8

19. A.A. Gusak, in *Handbook of Higher Mathematics*, ed. by A. A. Gusak, G. M. Gusak, (Navuka i technike, Minsk, 1991). 480 p
20. O.S. Grynchenko, Some applied models of strength reliability under sudden failures. Bulletin of the National Technical University "KhPI". Thematic issue: Dynamics and strength of machines, Kharkov, vol. 1(12) (2003), pp. 51–58
21. A.P. Prudnikov, in *Integrals and Series. Elementary Functions*, ed. by A. P. Prudnikov, Y. A. Brychkov, O. I. Marichev, (Nauka, Moscow, 1981). 800 p
22. Y.K. Konenkov, I.A. Ushakov, *Reliability of Electronic Equipment Under Mechanical Loads* (Sov. Radio, Moscow, 1975). 144 p
23. V.E. Kanarchuk, in *Machine Reliability: Textbook*, ed. by V. Y. Kanarchuk, S. K. Polyansky, M. M. Dmitriev, (Lybid, Kiev, 2003). 424 p
24. O.S. Grynchenko, Analysis of strength reliability of elements and systems based on the method of unit distributions. Problems of reliability of machines and means of mechanization of agricultural production. Bull KhNTU of Agriculture Named After Petro Vasylenko, Kharkiv **100**, 109–118 (2010)
25. E. Yanke, F. Emde, F. Lesch, *Special Functions* (Nauka, Moscow, 1977). 344 p
26. A.V. Sylvestrov, R.M. Shagimordinov, Fragile destruction of steel structures and ways to prevent it. Probl. Strength **5**, 88–94 (1972)
27. E. Dillon, C. Singh, *Engineering Methods for Providing Reliability of Systems* (Mir, Moscow, 1984). 318 p
28. DSTU 2860: 94. Reliability of technology. Terms and definitions. 88 p
29. A.I. Aristov, V.S. Borisenko, *Evaluation of Mechanical Systems Reliability* (Znanie, Moscow, 1972). 120 p
30. A.R. Rzhanitsyn, *Design Theory of Building Structures for Reliability* (Stroyizdat, Moscow, 1978). 239 p
31. V.V. Bolotin, *Application of Methods of Probability Theory and Reliability Theory in Calculations of Structures* (Stroyizdat, Moscow, 1971). 256 p
32. K. Kapur, L. Lamberson, *Reliability and Design of Systems* (Mir, Moscow, 1980). 604 p
33. I.B. Barsky, V.Y. Anilovich, G.M. Kutkov, *Tractor Dynamics* (Mashynostroenie, Moscow, 1973). 280 p
34. V.A. Zhovdak, I.V. Mishchenko, *Predicting Reliability of Structural Elements Taking into Account Technological and Operational Factors* (KhSPU, Kharkov, 1999). 120 p
35. N.N. Yatsenko, *Oscillations, Strength and Forced Tests of Trucks* (Mashinostroenie, Moscow, 1972), 372 p
36. A.S. Gusev, *Resistance to Fatigue and Survivability of Structures with Random Loads* (Mashinostroenie, Moscow, 1989). 248 p
37. O.S. Grynchenko, *Mechanical Reliability of Mobile Machines: Evaluation, Modeling, Control—X* (Virovets A.P. "Apostroph", Kharkiv, 2012). 259 p
38. E.S. Pereverzev, *Random Processes in Parametric Reliability Models* (Naukova Dumka, Kiev, 1987). 240 p
39. D.N. Reshetov, A.S. Ivanov, V.Z. Fadeev, *Machine Reliability* (Vyschaya Shkola, Moscow, 1986), 238 p
40. Y.B. Shor, F.I. Kuzmin, *Tables for Analysis and Control of Reliability* (Sovetskoe Radio, Moscow, 1968). 288 p
41. P.L. Vizir, Reliability of the system element. In the book: loads and reliability of building structures. Proc. TcNIISK, Moscow **21**, 26–42 (1973)
42. A.A. Kuznetsov, O.M. Alifanov, V.I. Vetrov, et al., *Probabilistic Characteristics of Strength of Aviation Materials and Assortment Sizes* (Mashynostroenie, Moscow, 1970). 576 p
43. V.Y. Anilovich, O.S. Grinchenko, V.V. Karabin, et al., in *Durability and Reliability of Machines*, ed. by V. Y. Anilovich, (Urozhay, Krasnodar, 1996). 288 p
44. V.A. Ostrejkovsky, *Multifactor Reliability Tests* (Energiya, Moscow, 1978). 152 p
45. E.S. Pereverzev, *Reliability and Testing of Technical Systems* (Naukova Dumka, Kiev, 1990). 328 p

46. M. Abramovitch, I. Stigan (eds.), *Handbook of Special Functions with Formulas, Graphs and Tables* (Nauka, Moscow, 1979). 832 p
47. B.V. Gnedenko, Y.K. Belyaev, A.D. Soloviev, *Mathematical Methods in Reliability Theory* (Nauka, Moscow, 1965). 524 p
48. I.B. Gertsbakh, K.B. Kordonskiy, *Failure Patterns* (Sov. Radio, Moscow, 1966). 168 p
49. A.S. Pronikov, *Parametric Reliability of Machines* (Publishing House MSTU Named After N.E. Bauman, Moscow, 2002). 560 p
50. N.L. Johnson, in *One-Dimensional Continuous Distributions [Electronic Resource]*, ed. by N. L. Johnson, S. Kots, N. Balakrishnan, (Binom, Moscow, 2014). 600 p
51. L.N. Bolshev, N.V. Smirnov, *Tables of Mathematical Statistics* (Nauka, Moscow, 1983). 416 p
52. V.Y. Anilovich, A.S. Grinchenko, V.L. Litvinenko, *Reliability of Machines in Tasks and Examples* (Oko, Kharkov, 2001). 320 p
53. V.Y. Anilovich, A.S. Grinchenko, V.L. Litvinenko, I.S. Chernyavsky, *Predicting Reliability of Tractors* (Mashinostroenie, Moscow, 1986). 224 p
54. G. Bateman, A. Erdei, *Higher Transcendental Functions* (Nauka, Moscow, 1974). 296 p
55. M. Lindbetter, G. Lindgren, H. Rotsen, *Extremes of Random Sequences and Processes* (Mir, Moscow, 1989). 392 p
56. E.S. Venttsel, L.A. Ovcharov, *Theory of Probability and Its Engineering Applications* (Nauka, Moscow, 1988). 480 p
57. O.A. Ryakhovsky, S.S. Ivanov, *Handbook of Couplings* (Polytechnika, 1991). 384 p
58. V.V. Nichke, *Reliability of Trailed and Mounted Equipment of Tractors* (Vuscha Shkola. Publishing House in Kharkov University, 1985). 152 p
59. O.S. Grynchenko, A.V. Zemnitsky, Influence of the parameters of safety couplings on the strength of machine elements reliability. Bull. KhNTU Agric. named after Petro Vasylenko **33**, 106–112 (2005)
60. O.S. Grynchenko, A.V. Zemnitsky, Experimental study of activation process of friction clutch. Sci. Bull. KhNTU named after Petro Vasylenko **41**, 381–386 (2005)
61. O.S. Grynchenko, Forecasting and reliability of machine elements in case of overload failures. Sci. Bull. Natl. Agrarian Univ **49**, 83–89 (2002)
62. O.S. Grynchenko, Models of reliability of elements and systems for protection against overloads with the help of safety devices, in *Technical Service of Agrarian and Industrial Complex, Technology in Agricultural Engineering: Bulletein of KhNTU Named After Petro Vasylenko, Kharkiv*, ed. by A.S. Grinchenko, vol. 76 (2009), pp. 155–161

The manufacturer's authorised representative in the EU is Springer Nature Customer Service Centre GmbH, Europaplatz 3, 69115 Heidelberg, Germany. If you have any concerns regarding our products, please contact ProductSafety@springernature.com

Printed and bound by CPI Group (UK) Ltd, Croydon, CR0 4YY
15/07/2026
02167652-0001